Reaching for Infinity: Puzzles, Paradoxes and Brainteasers #3

Reaching for Infinity: Puzzles, Paradoxes and Brainteasers #3

Stan Gibilisco

FIRST EDITION
FIRST PRINTING

Library of Congress Cataloging-in-Publication Data

Gibilisco, Stan.
Reaching for infinity: puzzles, paradoxes and brainteasers #3 /
Stan Gibilisco.
p. cm.
Includes bibliographical references.
ISBN 0-8306-8327-5 ISBN 0-8306-3327-8 (pbk.)
1. Mathematical recreations. 2. Infinite. I. Title.
QA95.G484 1990
793.7'4—dc20 90-10722
CIP

TAB BOOKS offers software for sale. For information and a catalog, please contact TAB Software Department, Blue Ridge Summit, PA 17294-0850.

Questions regarding the content of this book should be addressed to:

Reader Inquiry Branch
TAB BOOKS
Blue Ridge Summit, PA 17294-0214

Acquisitions Editor: Roland S. Phelps
Book Editor: Lisa A. Doyle
Production: Katherine Brown

For George, who ought to remember this
and
For Farah, in existence beyond reach

Contents

Introduction

SUGGESTING THAT THIS BOOK COVERS THE SUBJECT OF INFINITY IS SIMILAR TO SIMPLY saying that *aleph-null* is a representation of the Absolute Infinite. In other words, it is a skeletal work. A truly comprehensive treatise on the subject would require at least ten times this much space. Rather than serving as a landmark work on infinity (a subject that has interested me for practically all of my life), this book is intended to arouse curiosity and provoke thought.

Thoughts about the infinite have been around just about as long as there has been recorded history. The idea of the "unknowable" is a simple enough thing—simple, but not easy. It is like reaching out and picking the stars off the canopy of the celestial sphere. That would seem to be simple also, if only your arm was long enough.

But the arms of the human imagination have proven to be long enough to at least figuratively touch, if not actually grasp, those stars. We now know what the stars are, how they evolve, how they die, how hot they are, and how distant and numerous they are. All of this information was gathered from an observing post that resembles what an electron would be like compared with the room in which you now recline in a moment of time shorter than the instant needed for you to glance up and look at your watch.

If we can know so much about the universe by looking through such a tiny window for such a brief moment in cosmic time, how much more can we know if we employ the tools of the mathematician, which are instruments not constrained by the limits of distance, time, and dimension?

The history of the study of the infinite is replete with paradoxes and counterintuitive results. One is the so-called Berry Paradox, where it is first stated that there must be a smallest natural number too large to name, and then, having just described it as such, we refute the very statement of the theorem. Another strange consequence is Georg Cantor's discovery that many different "levels" or sizes of the infinite exist, the smallest being that of the natural or counting numbers, but greater and greater infinities being resolved without limit. The utter end of all this has been called the *Absolute Infinite*, and some of its properties have been speculated upon, but by its very nature it evades full and concrete grasp.

As there are infinitely large quantities, there must also be infinitely small ones, smaller than any positive real number but larger, in some sense, than zero. This subject is still largely unexplored. Such quantities have been called infinitesimals, and they are left without any apparent "use" and thus are neglected with the explanation that they are "sterile." That is too bad; exactly the same sort of thing was once said about the science of physics. This was around the end of the nineteenth century, just before the discovery of the special theory of relativity by Albert Einstein.

This book first explores the physical nature of infinity in the hierarchy of particles that exist in the cosmos. Then the various aspects of the infinite are explored as seen in number theory—arithmetic and algebra, set theory, and geometry. I also devise a coordinate plane that allows viewing the entire set of ordered-number pairs (x,y) at once within a finite area.

It has been said by some that infinity must exist because we can think about it, even though we cannot see it or describe it in its entirety. Others have suggested that there is no such thing as infinity in the real universe, and this is reflected by the very fact that we cannot count up to it. There wouldn't be room enough, say such "finitists," for an infinite number of anything in a finite space, and we can never, according to the theory of relativity, cover an infinite distance.

If you are a finitist (someone who does not believe in infinity), you might not accept much of the material herein.

In actual fact, it might not make any difference whether there is a real infinity or not! If there is not—let's just suppose there isn't—it sure is fun thinking there is. And even if infinity does not exist, there are finite quantities that are plenty too big for us to ever begin to imagine, as long as *we* remain finite . . . and we must remain finite, if you are a finitist.

There might be other universes and other lives outside of this transitory "threescore and ten." But no matter how large or how long-lived we ever find ourselves to be, as long as we are finite, there will be things much too great for us to fully comprehend. No matter how big something gets, there's room for more, even in a finitist's universe.

But what about getting sucked into a black hole? What happens after the universe dies out, either by growing cold and dark or by contracting to a single point, stopping time? According to finitist theory, neither of these things can ever take place. There would be a barrier, a point past which we could not go—and in order to define such a point, we imply that there must be something past it. The old "dart-at-the-end-of-the-universe" paradox tells us that. In this paradox, it was theorized that either there is an end to the universe, to space, or there isn't. Then a dart must keep on going forever, or it must stop at the end of space. But if it stops, there must be some object to stop it, some force, some barrier. How is this barrier made? If it exists, it must have some dimension—and this implies that there is something beyond the point at which the dart has stopped. So if you are a finitist, then you must accept that the universe will never come to an end . . . in other words, that it is infinite.

What good is infinity to us if we can never actually get there? We can imagine and evaluate it with all the tricks and tools of mathematics we can develop, but if we can't touch it, why bother with it?

The answer could lie somewhere in that curious thing we call the mind, but it is not up to me right now to try to find that answer. After all, what is "the mind," that it can think about things that might not be possible and still evaluate them and insist on pursuing knowledge that might never conquer a nation or bring riches and immortality?

I think of the mind as a window, a viewing port through which the brain is connected with the greater cosmos. We might put a telescope in the window to see things extremely distant. No matter how much magnification we allow this telescope, we still keep seeing more and more new things at greater and greater distances. We will, it seems, never run out of things to look at.

That is the whole concept of infinity: It's something we'll never get to entirely know.

Chapter 1

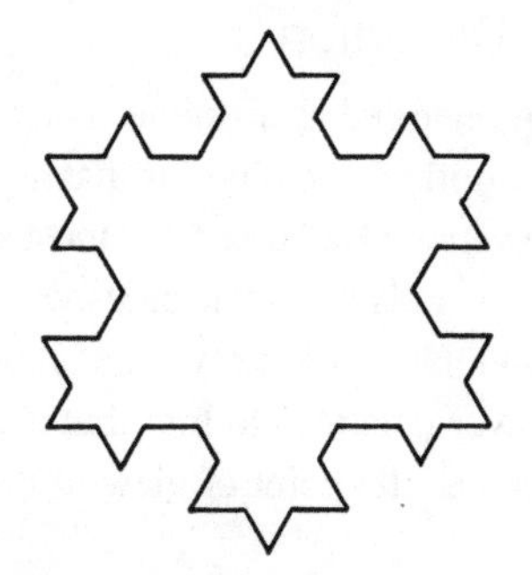

Infinity of Space and Time

MOST ARE TAUGHT THAT SPACE IS INFINITE, THAT IT GOES ON IN ALL DIRECTIONS without end. At least, this is what we learn initially. The idea that space might not be infinite—that there are only so many cubic miles or cubic light years of it—implies that there must be a boundary, a wall, a barrier. The schoolchild is assumed to be too naive to grasp a theory of a finite but unbounded universe. This theory is usually taught later in college, or more recently, in high school.

Likewise, most also perceive time as going on without end. This is probably because most cannot perceive of a beginning nor an end to all that is. If we were to locate an instant that could be verified as the very start of the cosmos or the originating time-point for everything to follow, we would ask, "But what came before that?" The answer, "Nothing came before that because there was no time before that," would grate on us and would not be acceptable. The same is even more true of an ending to all things: we cannot grasp our own mortality, let alone the mortality of a whole cosmos. Only later in life do many come to realize that time, too, might be finite without having bounds.

But which is ultimately the case? Are space and time truly infinite? Or are they finite but unbounded four-spheres and time-circles? Or are they finite with definite bounds? Even nowadays no one can say for certain which theory is actually correct, although we are coming to realize that infinity, if it indeed exists, is a complicated webwork of space-time fabric beyond the wildest science-fiction writer's dreams. As with so many aspects of nature, it is certainly true of space-time that reality is stranger than fiction.

The famous astrophysicist Stephen Hawking, reaching for the mind of the Creator despite physical handicaps, describes space and time as a "foam" of many dimensions, perhaps up to 11 different dimensions in all, with the three-dimensional case being special by virtue of its mathematical properties. His book, *A Brief History of Time*, has become a nationwide bestseller. It is testimony to people's curiosity, not only about the universe, but about the minds that strive to bring infinity within reach.

Einstein's Space-Time Distortion

Space and time were long regarded as absolute. Isaac Newton said that time "flowed along smoothly." This seems intuitively so obvious that any rendition of the principle must take the form of almost poetic verse. Of course time moves along smoothly. There are perhaps certain quantum leaps in periods of unconsciousness, or after death, or before birth, but certainly, time itself is inviolate. It simply *is*. It must, Newton thought, always go along at the same "rate." It never occurred to him that the rate of time progression might be influenced by anything, such as the point of view of the person observing (or perceiving) it.

This notion led into contradictory territory around the beginning of the twentieth century. At that time, physics was regarded as a science that had matured, as a pasture grazed bare of any further sustenance. All that was to be known, it was thought by some, was known. It was a lifeless pursuit. A few simple experiments were all that remained to seal the whole field of physics and allow it to be placed on the shelf as a work of art completed.

But those experiments proved to be a nuisance. They didn't turn out as expected. It was thought that visible light, and all forms of radiant energy, traveled through space via conduction in an omnipresent substance called luminiferous ether. This ether was supposed to conduct the vibrations of radiant energy, discovered not long before to have wave-like properties, just as air or water might conduct acoustic waves. This would result in a difference in the apparent speed of light arriving from stars in some directions, compared with other directions.

The motion of any given point on the surface of the earth is complex; it has components of the planet's rotation on its own axis, the revolution of the planet around the sun, and the motion of the sun through interstellar space (Fig. 1-1). And there are other components, too: the motion of the whole galaxy with respect to other galaxies, and the motion of the "Local Group" of galaxies relative to the universe as a whole. The luminiferous ether was thought to be fixed and stationary in some absolute sense. By measuring the speed of light from stars in various directions, it was thought that the absolute speed of the earth with respect to this ether might be found and, in that way, the absolute motion of any point of reference.

That didn't work out as planned. The speed of light appeared to be the same from all directions. It was a constant 186,282 miles (299,792 kilometers) per second, and the direction did not appear to have anything to do with it.

Explanations for this phenomenon were invented: perhaps the luminiferous ether was like the atmosphere of our planet and moved along with the earth, creating the apparent result that the speed of light was constant in all directions. But this kind of rationalization began to wear thin in the minds of even the most stubborn ether proponents. It was getting inelegant—and there is something objectionable about an inelegant theory. Something had to yield. It was up to Albert Einstein to change matters.

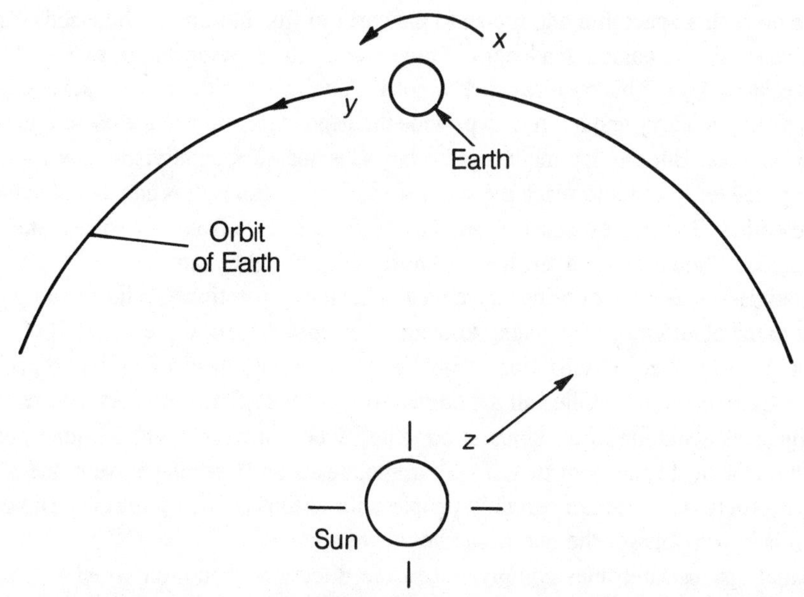

Fig. 1-1. The motion of any point on the surface of the earth has many components. Shown here are *x*, caused by the rotation of the planet, *y*, caused by the revolution of the planet around the sun, and *z*, caused by the movement of the sun through the galaxy. The view is from above the North Pole.

Einstein was a worker in the Swiss Patent Office, a malcontent in some ways because he was a person who did not seem to fit into the traditional academic mold. (In his later years, it is said that he regarded socks as superfluous; he was always simple and naive in certain formalities. He got straight to the heart of matters in ways that others didn't always appreciate.) Being grossly underemployed, he worked on his theory of relativity in the evenings, on weekends, or whenever he could get spare time. He realized that even a full 40-hour work week would account for less than half the waking hours of an ordinary person, and the rest of the time, except for necessities such as eating and washing, was his to explore the nature of the universe. He began with the premise that the speed of light, measured from any nonaccelerating point of reference, must always appear the same whether there was relative motion or not. It seemed somewhat against intuition to postulate this, but that is what experiments confirmed and most could not deny the actual empirical results of the experiments. It was true then, as it is now, that "one experimentalist can keep a dozen theorists busy."

From this concept of the constancy of the speed of light, it followed that space and time do not necessarily appear the same from different reference frames. Distances and time intervals can be made greater or less as a result of speeds of sufficient magnitude. This result might seem counterintuitive, as was Einstein's hypothesis to minds of his day, but the conclusions of the Special Theory of Relativity have been repeatedly verified by experiments since they were first published.

When observing events on a spaceship traveling at high speed, the rate of time on the spaceship appears to slow down. Thus, it might be possible for a person to travel a great distance on such a spaceship and return to the earth to find hundreds, thousands, or millions of years to have passed in a length of time that, to the traveler, might seem to be only a few weeks or days. This requires speeds not yet attainable with today's rocket engines; there is no way to carry enough fuel to provide the needed thrust, and there are other technical difficulties. But the day might not be far off when such relativistic space journeys become possible. In order to reach the stars, we will need spaceships capable of achieving these velocities. The nearest star is more than four light years from our solar system, and the Milky Way galaxy has a diameter of almost 100,000 light years.

Relativistic space travel brings up ethical and moral questions. Who would go on a journey to the other end of the galaxy knowing that upon return to the earth, 2000 centuries would have passed? Civilization would be far different. There might be no people at all; a holocaust could have killed all the humans off. Science-fiction writers have exploited this subject. Speculation of all kinds is possible. Such journeys might be quite possible even within the next century or two, if we extrapolate according to the current rate of technological progress. There are certainly people willing to risk such journeys. The ethical problem is how to answer the question: should we allow it?

Spatial distance becomes compressed in the direction of motion when a spaceship travels at extreme speed. This distortion of space is of little practical consequence except to demonstrate that the continuum can be "squashed." The factor for determining the extent of this spatial "squashing" is

$$L^* = L\sqrt{1 - v^2/c^2}$$

where L^* is the apparent length of the object as seen from a point of view "at rest," L is the actual length of the object "at rest," v is the speed of the object, and c is the speed of light in the same units as v.

As the speed of the object approaches c, the length L^* approaches zero. The time-dilation factor approaches infinity; that is, it increases without limit.

And there is another effect: the mass of the object also increases without limit in inverse proportion to the change in the length. This is one way of demonstrating that accelerating an object to the speed of light is not physically possible, at least not in the known universe. It would take an infinite amount of energy to do this. The mass would become so great at extreme speed, that further acceleration would be difficult. The total energy required to achieve a final speed of $v = c$, given any finite starting speed, can be shown to be infinite by evaluating an integral (the area under a curve).

At the speed of light, in theory at least, time would stop and spatial length would be reduced to zero.

If we could somehow accelerate to the speed of light, even for what to us would be just a short while, an infinite amount of time would pass in the outside universe. This is equivalent to saying that we would leave the space-time continuum as we know it. According to theory, this would be a "one-way membrane" that, once crossed, could not be gone through in the other direction. We could never get back to our cosmos.

We cannot generate sufficient energy to accelerate to the speed of light and reach infinity by throwing ourselves into the infinitely far-off future. We might go a thousand years ahead, or a million years, or a billion years, or any finite time until the universe collapses back on itself—if it ever will. But we cannot travel infinitely far into the future by this means.

By all physical methods, it appears that infinity is outside our grasp.

Space without End

Elementary geometry teaches the concepts of the point, line, and plane. A point is an infinitely small thing, with a precise location but no physical size. A line is infinitely long in two directions and infinitely thin, representing one dimension. A plane is an infinitely thin, flat membrane representing two dimensions. Since these are primary concepts, they are considered impossible to define rigorously, and no attempt is made to do so. We might just as well say that a line is "an infinitely thin thread that is pulled straight and goes on forever either way." Similarly, we might say that a plane is "an infinite pane of frosted glass without the glass." That mode of thinking is as good as any other.

The mathematician Euclid created a whole system of geometry based on the concepts of point, line, and plane. For this geometry to work, we need several postulates, or statements, taken as truth without proof. One of these postulates, changed a little bit in modern times, is called the Parallel Postulate. According to this axiom, given a line L and a point P not on L, there exists one and only one line M through P that is parallel to L (Fig. 1-2).

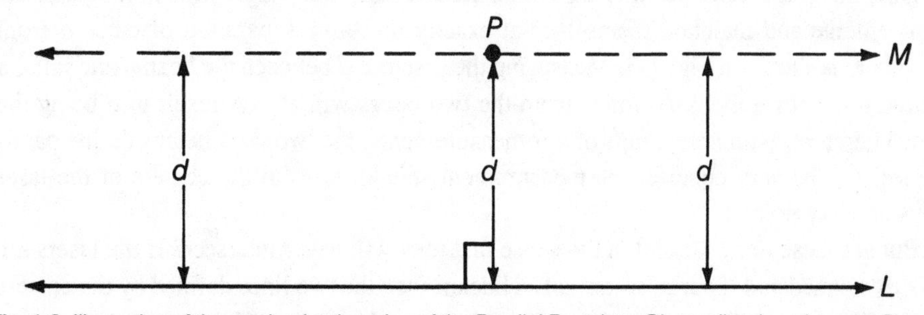

Fig. 1-2. Illustration of the modernized version of the Parallel Postulate. Given a line *L* and a point *P* not on *L*, there is one line *M* through *P* such that the distance *d* between *L* and *M* is the same everywhere—that is, the two lines are parallel.

This seems intuitively obvious. Two lines are parallel if and only if they lie in a common plane and they do not intersect. Skew lines may never intersect but they cannot lie in a common plane; any two nonparallel lines in the same plane must intersect at exactly one point unless they lie on the same line, and then they intersect at an infinite number of points.

But mathematicians found something unsettling about this postulate. It could be assumed, for example, that there might be no lines parallel to a given line through a point not on that line, and no contradiction would arise out of this assumption, even though it would make Euclid's Fifth Postulate—the equivalent of the Parallel Postulate—false. It is also possible to assume that there could be many lines, even infinitely many, that run through an external point P not on line L and yet are "parallel" to L in the sense that they are on the same "plane" and never intersect L.

Consider the surface of a sphere. To make this easier, imagine the surface of the earth or of a globe marked off with lines of latitude and longitude. How do you define a "straight" line—a true line—on this surface? If you take off in some direction that you consider "straight," you'd follow a so-called *great circle* or *geodesic path* over the earth's surface. An example is the route of a ship or airliner from San Francisco to Tokyo. The "straight" path is the shortest path between two points on the given surface. Any two non-identical great circles on the surface of the earth must intersect at two points. Therefore, there cannot be two "straight" lines that are parallel. They always intersect at exactly two points, and these points are exactly opposite each other on the globe. The meridians are a good example; they are all "straight" (great circles) and they all intersect at exactly two points, the North Pole and the South Pole.

When thinking of our universe and of space, we think of Euclidean space. That is, straight lines are really straight. But how do you tell if a given line is "straight?" The only mechanism available for that purpose is a beam of light. A beam of light always follows the shortest path between two points in a vacuum. (This is not true when light passes from one medium to another with different indexes of refraction.) A "straight" line can even be defined as a line followed by a photon in a vacuum.

Now imagine a laser beam setup with a second laser setup right next to it so that the beams emerge and maintain themselves at exactly the same separation distance *d* from each other, as shown in Fig. 1-3. Measuring the distance *d* between the beams one mile, a hundred miles, or a thousand miles from the two lasers will always result in *d* being the same. Therefore, within the limits of our measurements, the two laser beams define parallel lines. (To be very concise, the measurement should be from the centers of the laser beams at every point.)

But are these lines parallel in the sense that they will never intersect? If the lasers are perfectly aligned and if our universe is Euclidean, then the two lines defined by the centers of the laser beams will never intersect. But this cannot be verified. It is not possible to say that the lines will *never* intersect, because we cannot go on and on checking *forever*. Sooner or later something will limit us to prevent mathematically verifying this by experiment.

From a mathematical standpoint, then, we simply cannot say whether the two lines will ever intersect or not. We cannot even make this conclusion for laser beams that clearly diverge near their sources.

If the universe was like some huge four-dimensional sphere with a three-dimensional surface, then any two straight lines—lines determined by photons in vacuum—would always have two, and only two, points of intersection, assuming they were not the same exact line.

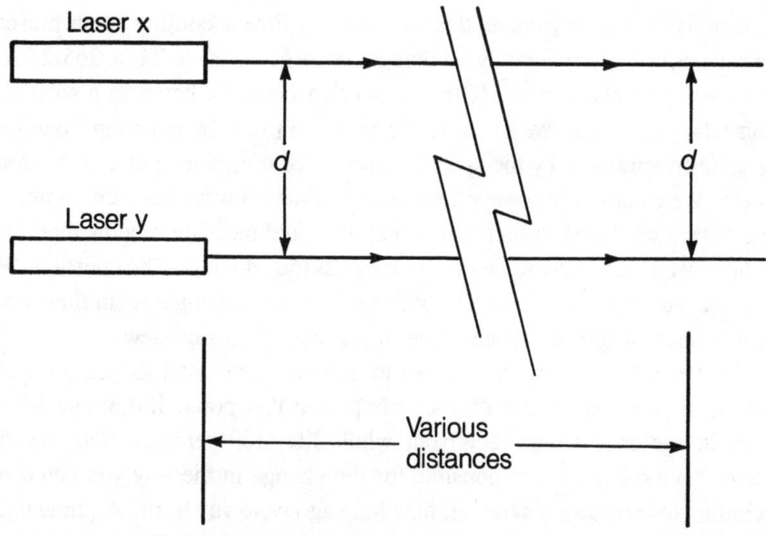

Fig. 1-3. Setting up two lasers so that the beams are parallel. The distance *d* between the centers of the beams is always the same, no matter what the distance from the lasers.

Modern astronomers have reason to believe that our universe is really shaped this way and that the four-sphere has a circumference ranging between 20 billion and 200 billion light years. Then all light beams actually move along geodesics on this four-sphere, and, if allowed to travel for long enough unobstructed, they will eventually intersect at some common point, and then at a second common point, 10 to 100 billion light years further along.

It is believed by many cosmologists that space is finite but unbounded in the same way that the surface of the earth is finite without boundaries. Then space has only a certain number of cubic light years. That number is extremely large, but not infinite. The difference between any number and "infinity" is like the difference between apples and oranges. The two concepts are not at all the same.

Time without End

It is difficult to imagine space that is finite and yet unbounded because most cannot directly envision a four-dimensional sphere. We have only our perception of the globe or the earth's surface and our faith in mathematics to guide us in conceiving the geometry of four dimensions. The surface volume of a four-sphere can be calculated as can the total interior four-space volume of the four-sphere.

What of time? Is time infinite? Most cosmologists do not think so. The popular theory nowadays is the so-called Big Bang Theory, so named because the universe supposedly exploded from a primordial fireball some 10 to 20 billion years ago. Before that time, the equations break down; they do not give us any indication of the universe prior to that instant. The idea of a creation, with nothing before then, appeals to theologists who might say "We have been telling you that for centuries." Indeed, the Biblical account of the Creation and the scientific description of the Big Bang have many similarities.

But one thing is missing from all this: we measure time according to our present standards. A year might more accurately be thought of as $60 \times 60 \times 24 \times 365.25$ seconds: there are 60 seconds in a minute, 60 minutes in an hour, 24 hours in a solar day, and 365.25 solar days in a year. We measure the second in turn by counting "one thousand one" (the primitive way) or by the specific number of oscillations in certain atoms (the modern way). We cannot say, however, that seconds have always been the same. Perhaps atoms have always oscillated in such a way that we could measure time by them, but they might not have oscillated at the same rate long ago as they do now. The question "Relative to what?" always comes into discussions of this sort, but what might seem finite according to certain references might not appear finite from other points of view.

Take, for example, the way time seems to go faster and faster as you grow older. A year at age 50 is just two percent of your lifespan to that point. But at age 10, it is 10 percent, and at age one, it represents your whole life (100 percent). Suppose this—the experience of having lived—is responsible for the change in the way you perceive time. Then, according to your own reckoning, how long ago were you born? A "time unit" may be taken as one percent of your life. This choice is arbitrary, but it does not matter as long as this same standard is kept throughout. When you are 50-years old, one percent of your life—one time unit—is a half year. When you are 10-years old, it is 0.1 years or about 36.525 days. When you are a day old, one time unit is 0.24 hours. All of these are seen according to the "present" or "initial" value representing your age.

Extrapolating back still further, when you are a minute old, one time unit is just 0.6 seconds. Here the question arises: when does the count begin? Let us play it safe and begin counting at conception. (It doesn't matter; it can be taken at any point between conception and birth in this model and the result will be the same.) One second after conception, a time unit would be just 0.01 second or 10 milliseconds. We could continue this backward in smaller and smaller increments with no end to it. The time units just get smaller and smaller.

Hence, how many time units old are you? The answer is: an infinite number! If you perceive your age in time units, accounting for the illusion that time goes faster and faster as you get older (since a day is a smaller and smaller part of your time unit as you age), then you perceive yourself as having been conceived infinitely long ago. No wonder you don't recall the event!

This might seem silly, but suppose the aging of the universe takes place in this way, so that the "second" that was used as our "time unit" might in fact be compressed long ago so that there have been infinitely many of them since the Creation. You then might say that the creation really never happened—that it is lost in the memory of the universe.

It appears that the Big Bang Theory demonstrates that there actually was an explosion a finite number of seconds ago, but it could depend on how you define the unit called a "second." Is it the time light requires to go 186,282 miles? Or the number of oscillations of a certain atom? Or is a second a certain fixed fraction of the age of the universe? In this last case, the universe was necessarily created an infinite number of seconds ago, applying the same logic as with the example concerning your perceived age.

Looking toward the future, will the universe ever collapse again? If so, might "time units" compress again so that there will be an infinite number of them towards the end? According to one model by Stephen Hawking where time actually goes backward in a shrinking universe, such a model might be consistent. But this is speculation, more or less. It might well be that the end will come in a finite amount of time and that we will simply not live long enough to see it.

Theories of an oscillating universe give some hope to those who want to explain things before the Creation and after the great collapse or implosion (if indeed there ever will be one). This could lend credence to the notion that time, like a Euclidean line, goes infinitely in both directions.

But if space can be non-Euclidean, cannot time also be that way? The answer is: possibly. Then we might have finite and yet unbounded time: a time circle, in which events are repeated again and again, exactly as they occurred the previous time around. In an oscillating universe, this could be equivalent to saying that the previous universe, and the next one, in fact all the repeated universes, would be identical with the present one. Not only would the maximum size and the duration be the same, but everything else also—including the existence of you and me. We would not remember having existed before, but we would have, in the sense that time would have come to the same point in the circle again. In this way, time might be finite, there being only a certain number of "years" or "seconds." Nevertheless, there would not be a boundary: no beginning and no end.

This is a difficult and somewhat haunting concept, but mathematically it is easy to explain. It could conceivably be physically real.

Hyperuniverses

Might there be universes of four or more spatial dimensions? This is another concept that is difficult to directly envision but not hard to deal with in the mathematical sense. Simple facts can be derived without much trouble about objects in four or more dimensions.

Consider the diagonal of a unit cube. A unit cube measures one unit of distance along each edge. There are six faces to a cube; each face of a unit cube has one square unit of surface area and the whole cube displaces one cubic unit of volume. The diagonal of the cube can be shown to be exactly $\sqrt{3}$ units. The simplest way to find this diagonal is to first find the length of the diagonal of the base of the cube—the diagonal of a unit square and hence a line segment measuring $\sqrt{2}$ units—and then, employing the Theorem of Pythagoras, find the length of the hypotenuse of a right triangle having the diagonal of the cube as its longest side (Fig. 1-4).

Continue to employ the same method for hypercubes of more than three dimensions. For the four-cube or *tesseract*, the diagonal would measure $\sqrt{4}$, or exactly 2, units if each edge is one unit long. In fact, it can be shown by mathematical induction that for n dimensions, where n is any integer greater than zero, the diagonal of an n-cube is exactly $\sqrt{n}$ units when the edges each measure one unit. (The case in one dimension is trivial. It is a line segment with a length of one unit.)

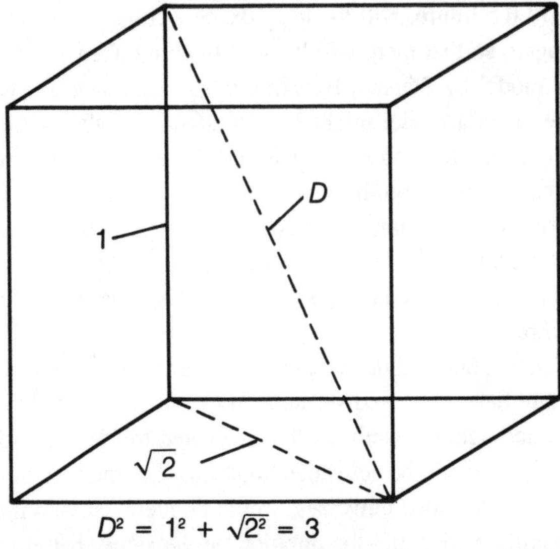

Fig. 1-4. Finding the diagonal of a cube using the Pythagorean Theorem. The base is the diagonal of a square; the height is the edge of the cube. Then the diagonal of the cube is the hypotenuse of a right triangle, as shown.

In four-space, it would be possible to place four sticks together at a common point and have them all be perpendicular to each other at that point. This, most cannot envision. (There are some mathematicians who say that they can begin to envision this after years of practice. However, if you practice in such thoughts for years, you might be likely to see almost anything.) In a universe with n spatial dimensions, you could do this with exactly n sticks. The sticks might be thought of as lying along the axes of a Cartesian system of coordinates. In n dimensions, it takes exactly n coordinates to uniquely define the location of a point. One type of coordinate system is the Cartesian system, which has mutually perpendicular axes where each axis is marked off in units of equal size, both positively and negatively, and intersect at the origin point $(x_1,x_2,x_3,\ldots,x_n) = (0,0,0,\ldots,0)$.

In an infinite space, these axes would all extend forever in both directions. But in a non-Euclidean space resembling a sphere, the axes would all reconverge at some distant point, a second origin. The two origins would be interchangeable; both of them would be composed of n axes, all mutually perpendicular. The maxima for the axes might be $\pm m$, where m would be the number of units in exactly half of the circumference of the hypersphere.

Do such hyperuniverses actually exist? According to Stephen Hawking, yes: but the three-space case is special. It is in this number of dimensions that we exist, and, to Hawking, this is no accident. In a two-space, the alimentary canal of a creature would cut the creature into separate halves, and circulatory systems as we know them would not be physically possible. In four dimensions, the inverse-cube law would hold for gravitation, rather than the inverse-square principle as in three dimensions. This would make it highly improbable that a planet could ever maintain an orbit around a parent star—one of the prerequisites for life as we think of it.

Nonetheless, the curvature of three-space, the fact that it is non-Euclidean, implies the existence of a fourth spatial dimension. We cannot have a non-Euclidean n-space without the existence of at least $n + 1$ spatial dimensions. The three-space in which we exist has been demonstrated to be non-Euclidean: gravitation and acceleration cause photons to move in paths that, although representing the shortest distance between two points, are still not straight lines. The most familiar astronomical example is the "bending" of light beams from distant stars, when that light passes very close to the sun (Fig. 1-5).

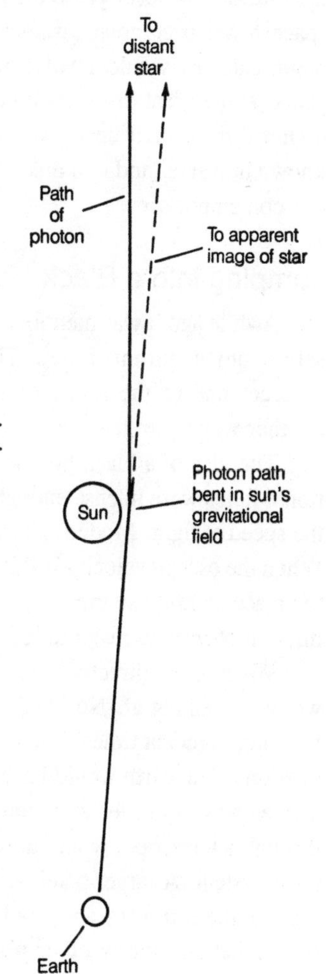

Fig. 1-5. The apparent direction of a star is affected by the proximity of the sun to the rays of light from that star. This is because gravitation bends the light beams from the star.

How many dimensions can there be? In theory there is no limit. Hawking suspects there may be up to 11 dimensions, with three-space being "spread out" and higher-dimensional hyperspaces "tightly curled up." We might think that every dimensional space is non-Euclidean in the space of one more dimension; there are infinitely many distinct ways a space might be non-Euclidean, but only one way it can be Euclidean. This gives rise to the theory that, as measured in progressively more spatial dimensions, the shortest path between two points gets shorter and shorter. If the distance in three dimensions, for example, is denoted by d_3, then the distance d_4 in four dimensions is less than d_3; the distance d_5, in five dimensions, would be less than d_4. In general, the distance d_{n+1}, in $n + 1$ dimensions, would be less than d_n in n dimensions, for all integers n.

If this is the case, then to what limit will the d_i converge? Will there be some lowest finite value, still more than zero—or will the limit be a single point in an infinite number of dimensions? It does not matter to us from a practical sense right now, perhaps, but in the future, if we ever get to the level of sophistication where we can use "wormholes" for space travel over great distances, it might be very significant. In three-space, going to other galaxies would involve relativistic time displacements of millions or billions of years. If the great distances could somehow be made much smaller by means of entering higher-dimensional universes, we might overcome this handicap and travel all over the known universe (and into unknown ones, too) and be able to come back and tell about it to our contemporaries.

Jumping into a Black Hole

Awhile ago it was mentioned that we cannot, using known physical means, throw ourselves into the infinite future. This was based on the notion that it is physically impossible to accelerate to the speed of light, making time dilation infinite. But there might be another way to stretch time infinitely; jumping into a black hole.

The idea of a black hole is not all that new. Long ago, it was thought that if gravitational fields were intense enough, the escape velocity might be as great as, or greater than, the speed of light. Modern physics has shown that this is possible, at least mathematically. When the escape velocity is this great, time comes to a complete standstill. It is equivalent to an acceleration so great that even light beams cannot keep up with it. The result is time dilation of infinite magnitude.

We have not directly observed a black hole and known with certainty that that is what we were looking at. No black holes have ever come wandering into the solar system, at least not in recent times. Even a small black hole would have profound gravitational repercussions. The earth would have to be about the size of a grape, but have the same mass as it does now, in order to become a black hole; a grape-sized black hole, while invisible through a telescope, would upset the orbits of all the planets as it came careening into the solar system. A larger black hole, having a radius of, say, 2.9 kilometers, would have the mass of the entire sun and would cause a catastrophe that would probably obliterate life on our planet and throw other planets into wildly eccentric elliptical orbits. Some planets would probably be ejected from the solar system altogether. Even a black hole of that mass is tiny by the standards of the universe. Much larger black holes are probably the general rule: they would have masses of millions or billions, or even trillions of suns.

The larger a black hole becomes, the lower its density. Small black holes would crush us if we jumped into them. Larger ones might not. It is this possibility that brings up the realm of infinite time travel. Crossing the "event horizon"—the point at which all things, even photons, are captured by gravity—would deposit us into the "infinite future," another time line from the one we know, separate from the one we know. This is, in fact, as we shall later see, one way of looking at infinity: when we go off to the infinite along one line, we actually put ourselves onto another line, where we can begin all over again.

There is a theory that the entire universe as we see it is one gigantic black hole, perhaps attached via an "event horizon" to another universe of the same number of dimensions, another bubble stuck to the one we find ourselves in.

It is thought that the entrance to a black hole is a one-way membrane. That is, it can be passed through in just one direction. It is not specifically known whether, once having entered a black hole, it is possible to ever come back out to the original universe. Mathematically, the models seem to say that we would be forever removed from our universe if we crossed such a one-way membrane. Thus, the model of bubbles stuck to each other is an imperfect one, a mental crutch that does not take everything into account. In this sense, it is a typical crutch we must use when pondering the infinite, the polydimensional.

What Is an Instant?

One popular model of time, and a familiar one to history students, is the line along which the "present" moves as a point traversing the years, decades, generations, centuries, and eons. These time lines grace the pages of all manner of books and are found superposed above the blackboards of many school classrooms. We may divide time into three distinct subsets: past, present, and future. The past and the future take the form of geometric rays (Fig. 1-6), and the present is a single geometric point.

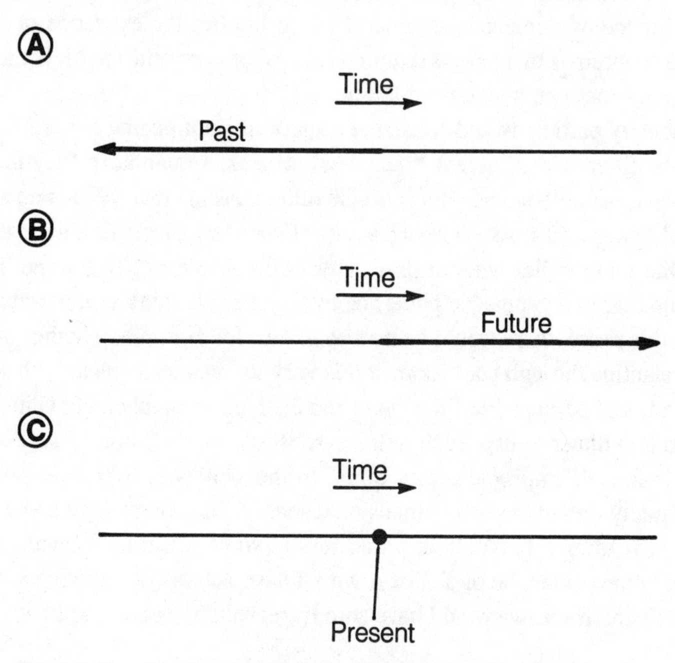

Fig. 1-6. Time-line illustrations of the past (A), future (B), and present (C).

But is the present actually a point with no duration at all? If so, how can we perceive it? The question is probably of no practical consequence, being in that realm of Hardyist thinking that has no "useful" application. (This term comes from the mathematician G. H. Hardy's name. He was a pure mathematician who defended his positions in a book called *A Mathematician's Apology*.) But if the present is not a point, how long does it last? A few milliseconds? A few minutes? The time it takes for a photon to go from one end of an elementary particle to another? The time required for a photon to travel its own length through space?

In a mathematical sense, a point is infinitely small. No mater how tiny a spatial displacement we specify, a point is smaller. We could stack any number of points on the head of a pin, for example, and always have room for plenty more. A point can be divided into any number of subpoints, all of the exact same dimensions as the original point. Perhaps the best way to look at the nature of a point is to regard it as a set of coordinates in a system of coordinate values, such as Cartesian three-space or spherical coordinates. A time point would have no duration. Our perception of the present, in contrast, must have duration sufficient for our brains to carry out the cognitive process of awareness. That is perhaps a moderate fraction of one second. It is a practical "instant."

The idea of quantized space-time occasionally arises in physics. Perhaps in the real universe, there is a smallest unit of distance that cannot be further broken down. This might be the size of an elementary particle, if such a thing actually exists. Whatever it is, it is very tiny, and we have yet to see one. If we ever develop a microscope that is powerful enough to resolve space down into elementary units, we will know it. We will be able to see things as a mosaic that further magnification does not render any more detailed (Fig. 1-7). The existence of elementary quanta of space implies the existence of time quanta, also: the time quantum is that period required for a photon or other particle traveling at the speed of light to cross one quantum of space.

An elementary particle would consist of a space quantum being either filled or empty. Space could be filled with different "densities" of matter quanta, or they might all have the same density, being identical. But it is also quite possible that there is no such thing as a quantum of space and no such thing as an elementary particle. Then, things would become smaller and smaller without limit, always breakable into littler and littler pieces.

When thinking of a geometric plane, or two-space, we think of a membrane without thickness. In this sense, there would be no way to tell, for example, whether or not there is a "flatland" slanting through your room at this very moment, complete with solar systems of its own kind, and perhaps life (in spite of the difficulties Stephen Hawking has pointed out for life in two dimensions). Such a universe would be so "thin" that it would not be there in any practical, empirical sense, to us. In the same way, our three-space universe would be infinitely thin in the four-dimensional sense, and if there were some four-dimensional creature reading a book about paradoxes in some tesseract-shaped room through which our universe passes, he or she or it would have no way of knowing we were there. Nor would we perceive it: we would have zero hypervolume in common with its universe.

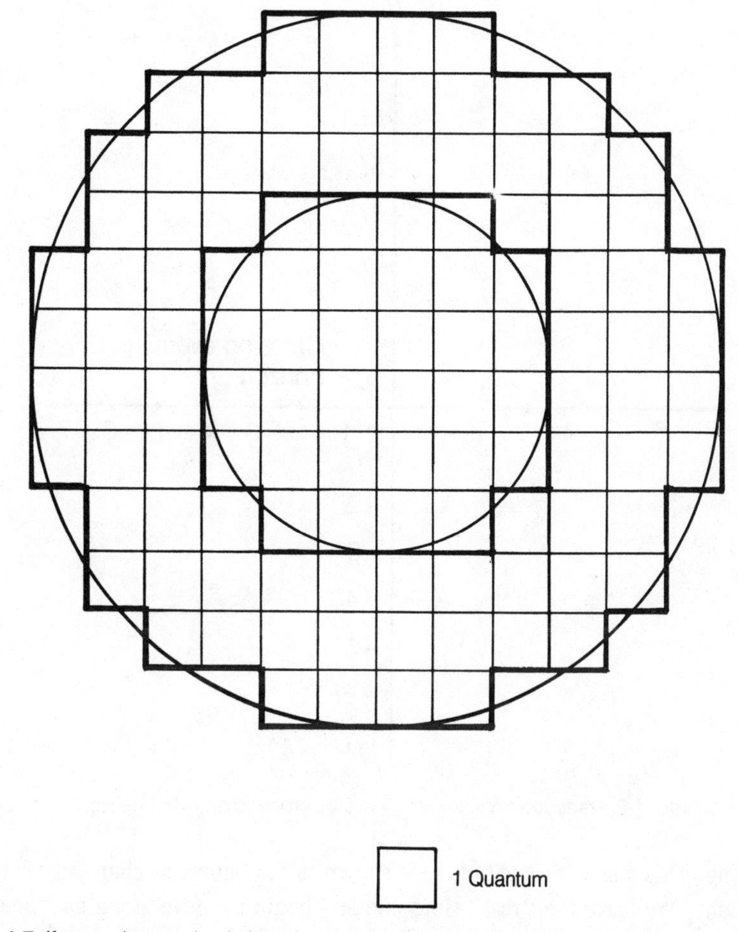

Fig. 1-7. If space is quantized, there is a point beyond which further resolution is impossible.

But perhaps if space and time are quantized, the situation is not so simple. Then a continuum might actually have some finite, nonzero "thickness" relative to the next higher-dimensioned universe. In such a circumstance, then, we might be able to detect that two-space slanting through the room; the four-space beings (if they exist) might be able to detect the presence of our universe. Perhaps most significant, our three-space would have finite, nonzero "thickness" in four dimensions and more—at least one space quantum thickness. This might be a very tiny spatial value, but it would be nonzero.

The point, or quantum, of time that represents the present in the strictest mathematical sense might move from one time line to another if we were to jump into a black hole. It is conceivable that there is more than one single dimension of time, more than one fixed "direction" in which time can go. We could even illustrate this as shown in Fig. 1-8: Jumping into a black hole, we take off along a time line perpendicular to the one we were

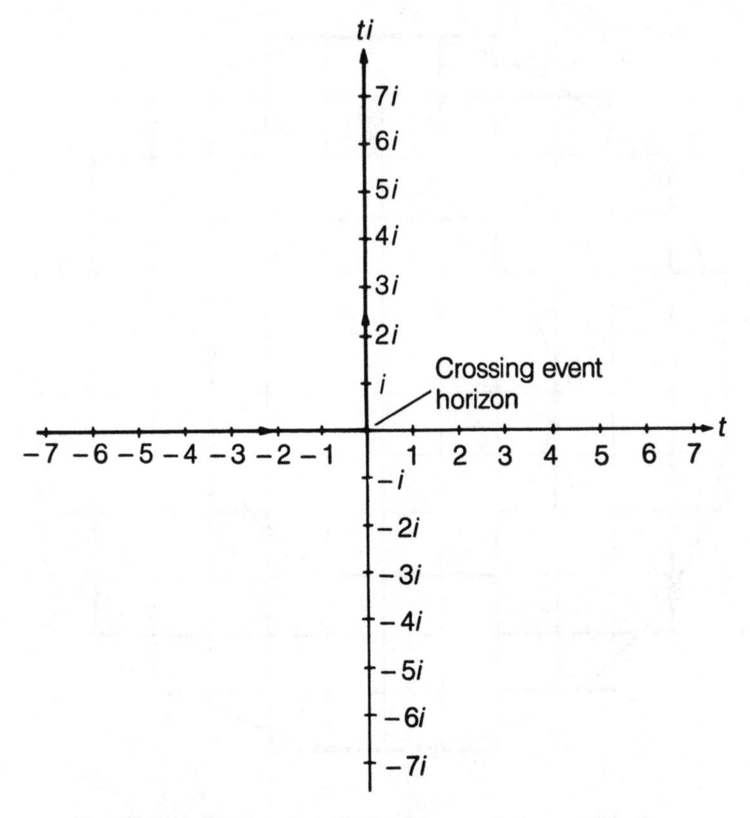

Fig. 1-8. Transition from time *t* to line *ti* by crossing an event horizon.

moving along. This can be thought of, as is shown in the figure, as changing vectors on a complex plane. We leave the "real" time line and begin to move along an "imaginary" time line, the time quanta now being in multiples of the elementary imaginary number $i = \sqrt{-1}$ rather than multiples of the real unit 1.

This might occur more than once. There might be a third axis as shown in Fig. 1-9 so that we may jump into a black hole once, going from the "real" time line (multiples of 1) to the "imaginary" time line (multiples of i) to yet another "super-imaginary" time line with its time units in multiples of some as-yet undefined number—call it I. This could occur over and over again, as many times as we might like. We could have hundreds of different time dimensions.

In some physics models of space-time, entering a black hole does indeed cause time to become imaginary. This might seem like a cop-out: we can always get away from dealing with the infinite hereafter by assigning a new time line or new time dimension to events that take place inside event horizons. But this could in fact be what happens. Perhaps someday we will be able to go through such an event horizon to find out, and in some way, communicate our findings back to our contemporaries.

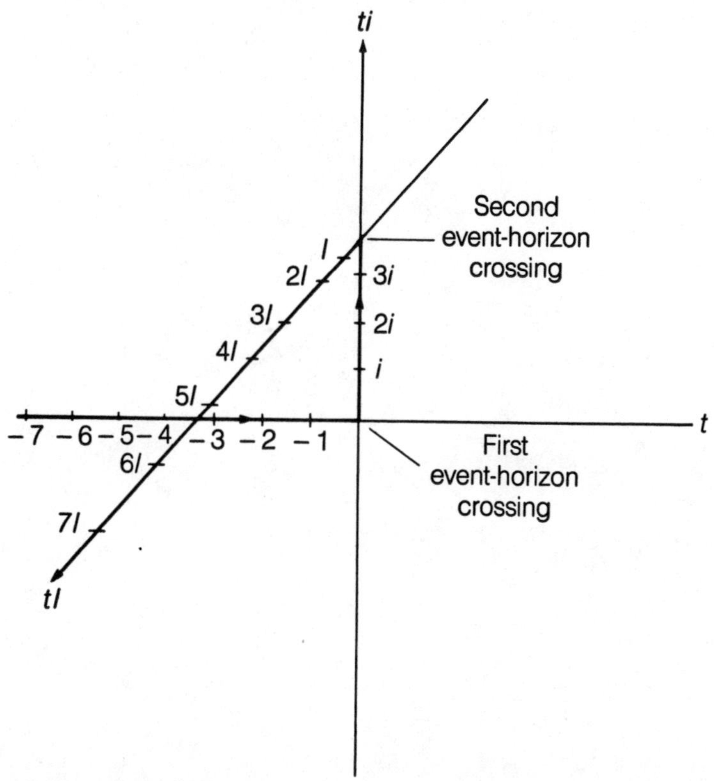

Fig. 1-9. Two time-line transitions. First, from time line *t* to line *ti*; second, from time line *ti* to line *tl*. Both transitions might occur because of passage through event horizons such as black holes.

Chapter 2

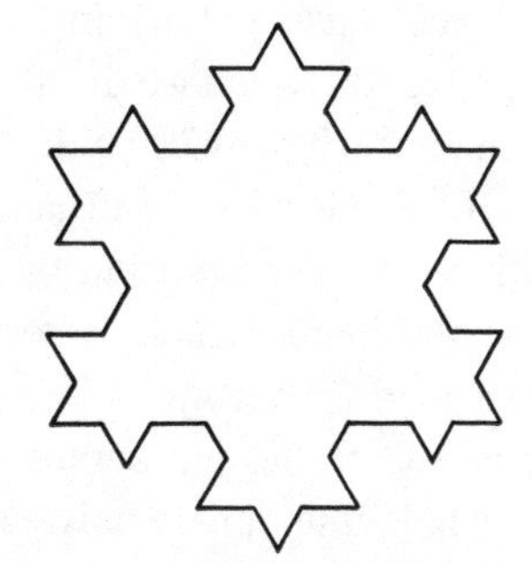

Particles without End

SOME SCIENTISTS HAVE POSTULATED THAT THERE MIGHT BE NO SUCH THING AS A fundamental particle of matter and that the structure of materials can be broken down further and further into smaller and smaller constituents, and we will be limited only by the precision with which we can make observations. It could well be that this is true, but it is doubtful that anyone will ever be able to offer a conclusive mathematical proof of it. And it could be false. Perhaps there is indeed a fundamental particle of matter, either because space itself is quantized or because there simply is a unique type of particle that cannot be further divided. No one will very likely come up with a rigorous mathematical proof of that, either. Physics is not like mathematics when universes are compared. The mathematician G. H. Hardy pointed this out: he claimed that given a mathematician and a physicist, the mathematician would be more in contact with reality because mathematical universes can be precisely explored while physical ones cannot. The set of oranges in a basket is a perfect concept. The structure of the actual oranges, in the physicist's eye, is a countless collection of whirling particles having this or that charge, this or that force acting on each other, emitting various wave-like, particle-like packets of energy. The mathematician disregards all this and considers the mathematics of the set. It could just as well be a set of reasons to be happy, as far as he cares.

Paradoxes and puzzles are usually thought of as mathematical in their nature. But the physical universe contains puzzles of its own. The hierarchy of particles might not be endless, but it is certainly complicated. It was Albert Einstein who said, "God may be sophisticated, but He is not malicious." Presumably that meant that the universe, while complicated, is not that way out of spite. God doesn't want to make it difficult for us. If He did, however, He probably couldn't have come up with a much more complicated puzzle than the true nature of the structure of matter.

The Largest Particle Known

For us, the largest "particle" that we can know is our entire universe. The universe is thought by many astronomers to have the form of a sphere in four dimensions: a set of points nearly equidistant from some center point P in four spatial dimensions. Thus the four-sphere has a three-space surface, on which we dwell. This sphere is thought to have a circumference of from 20 billion (2×10^{10}) to 200 billion (2×10^{11}) light years. A light year is about six trillion (6×10^{12}) miles, or the distance that a ray of light will travel through space in one year. This theory derives from the general theory of relativity, in which three-space is found to be non-Euclidean. A truly straight line does not really exist in such a space. Any line, if extended far enough, will eventually curve back on itself.

The well-known astronomer Carl Sagan, author of *Cosmos* and other books, describes a theory in which the whole four-sphere universe is actually an *elementary particle* in some much larger cosmos. It would have to be much larger: perhaps hundreds or even thousands of orders of magnitude larger, and in another spatial dimension as well. We cannot see anything of this place, if there really is such a place, nor can we surmise much about the way the laws of physics might operate there. The theory is essentially speculation. Equally speculative is the idea that perhaps there are elementary particles in our universe, and they are each universes unto themselves, three-spheres with two-space surfaces.

Stephen Hawking has pointed out that there would be difficulties with physics as we know it in universes with fewer than three dimensions or more than three. In a two-space, a creature would be split into two separate parts by its alimentary canal. In four-space, gravitational forces would operate according to a law of inverse cubes, rather than the law of inverse squares, as it does in three dimensions. Therefore, planets would have to be in almost perfectly circular orbits or they would either fall into their sun or fly off into interstellar space. Keep in mind that these observations presuppose that life in two-space would bear resemblance to life in three-space as we know it; it also supposes that the structure of a universe must involve stars and planets in order to contain anything meaningful. Neither of these presuppositions need be true. Things could be unimaginably different from anything mortals have ever dreamed up. Matter and energy, and time and space, could be completely unlike the things we call by these names in other universes. Therefore, intelligence could develop in ways we cannot imagine.

The material universe, as we know it and envision it, is the largest of the hierarchy of particles in the known cosmos. It has, according to modern thought, not always been as large as it is now. In fact, there was a time when it was small, perhaps even smaller than whatever elementary particle we might ultimately find. About 10 to 20 billion years ago, it is believed, the matter in the universe "blew up" from a particle given the name *ylem* (eye-lem) by the cosmologist George Gamow. This "Big Bang Theory" is supported by several observations, such as the red shift of light from far-off galaxies and the radiation that comes from every direction, presumably from the original hot "fireball."

The largest known particle is a changing entity.

Quasars and Galaxies

The next largest particle in the universe is the galaxy or quasar. These objects are roughly comparable in mass and luminosity. Galaxies appear to exist in many different shapes and over a wide range of sizes. Spirals, ellipsoids, and irregular galaxies abound.

All galaxies consist of lumps of matter that individually account for most of the radiation. These are the stars. There is also an abundance of hydrogen gas and interstellar dust. No two galaxies are exactly alike. The galaxies, like the larger universe, are evolving as their individual stars are born, spend their energy by atomic fusion, and die. Some of the mass in a galaxy is not visible through earthbound telescopes, because these objects, such as the planets and asteroids, do not give off light of their own. Then, too, a portion of the mass in a galaxy could be in the form of black holes—matter that has collapsed under the force of its own gravitational field. Our own sun is one of about 100 billion (10^{11}) stars in our galaxy, known as the Milky Way because of its appearance on a moonless night. Our galaxy is a spiral type, and we are located about three-quarters of the way from the center to the edge.

Quasars have been discovered fairly recently and are all extremely far away. The nearest ones are so distant that they have significant Doppler red shifts in their spectra. This has led astronomers to believe that quasars, because of their great distance and resulting time displacement, might be protogalaxies. If a given quasar is five billion light years away, we see it as it was five billion years ago. This could, in fact, be the way most or all galaxies appeared at that time. The universe was only about half to three-quarters as old then as it is now. There is still some doubt as to exactly what quasars actually are and what they look like close up. Some of them seem to be spewing out large amounts of matter, and all of them emit tremendous quantities of electromagnetic radiation at all wavelengths.

Some galaxies could be invisible, having fallen into their centers because of gravitation. In such cases, we might have no way of knowing from our vantage point that they exist at all. If there is sufficient mass in the universe, cosmologists predict it will finally stop expanding and begin to collapse because of the gravitation of all the matter in it. At present, the estimates of the mass of the universe are too small to allow this eventual contraction to occur. There appears to be only about one-tenth the needed amount of matter for this to happen. However, it could be that the remaining matter has not yet been detected because it is in the form of black holes. Scientists would rather (for the most part) believe in an unending cosmos, perhaps an oscillating universe, because it is more appealing to our innate desire for immortality. However, if present estimates are anywhere near correct, the universe will continue to expand until all the energy from atomic fusion in all the stars has been spent and re-spent, leaving a cold, dark, still-expanding multitude of elements and compounds.

Some cosmologists think that there is a large black hole at the center of our own Milky Way galaxy. The concentration of stars there is much greater than it is in the vicinity of our solar system. In fact, it seems that spiral galaxies are swirling cosmic hurricanes with gravitation greatest at the center, and if enough stars were drawn closely together enough, they would collapse from this gravitation and withdraw from the universe as we know it: they would become a massive black hole. Then our galaxy could be more massive than we think. This error could not be very great, however, because the mass can be estimated quite accurately according to the time it takes our sun to complete one revolution around the galactic core. This core is about 40,000 light years away.

Black Holes

What, exactly, is a black hole?

Any object with mass also has a gravitational field. This gravitational field might be comparatively weak, such as that surrounding asteroids or small planetary moons. On the earth's surface, gravitation is what we are used to considering "normal" and has the intensity equivalent to an acceleration of about 9.8 meters per second per second (m/sec^2). But it can become far more intense, to the extent that the escape velocity could exceed the speed of light. The escape velocity at the surface of our planet is about 25,000 miles per hour (11,000 meters per second). The speed of light is 186,282 miles per second (299,792 kilometers per second).

If the escape velocity of an object was greater than the speed of light, we would see nothing radiated from it. It would look black, and because of its propensity for drawing things in and never letting them come back out, it would behave just like a hole. But unlike a hole in the ground, which fills up, a black hole can never be filled up no matter how much matter is pulled into it. It simply gets larger and larger as it draws in more mass. A very small black hole, say, one the size of a grape, could suck in the whole planet earth and become only twice its former size. In theory, a black hole of this type might eventually devour the entire sun if there were a collision between our parent star and the black hole at a slow enough speed. It would then grow to a radius of about 2.9 kilometers, still not very large, since there may exist black holes as large as the whole cosmos we know. According to one theory, in fact, our universe is actually a gigantic black hole, and this is why it is closed off onto itself.

Scientists think that perhaps the object Cygnus X-1 is a black hole. It is invisible, but the effects it has on the motions of a nearby star indicate that it has considerable mass. Of course, this object could be a collapsed star that has not actually reached the black-hole stage.

The more massive an object, the larger the black hole it will create if it collapses under the influence of its gravitation. Also, the more the mass, the less the density of the actual black hole. A person diving into a black hole massing the same as the earth and being just a centimeter or so across would be crushed beyond recognition. The same is true of a sun-sized black hole. But if a whole galaxy were to fall into itself, the resulting density might be low enough to allow spacecraft to enter without being crushed to ooze.

Black holes cause space to become curved tremendously, to "pucker" in a formation shaped like a four-dimensional funnel (Fig. 2-1). I call these warpings of space "ngaldi" formations. I was not able to find any other single-word name for them. All masses, whether they are black holes or not, are surrounded by ngaldi formations, but black holes

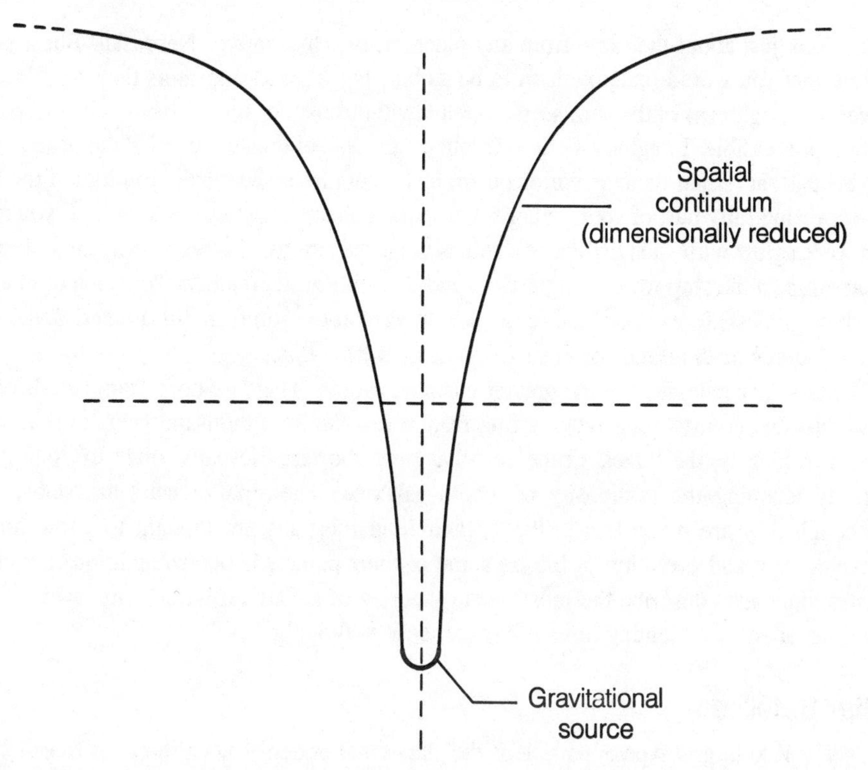

Fig. 2-1. Curvature of space in the vicinity of an intense gravitational source. This drawing is dimensionally reduced.

are especially noteworthy because of the extreme distortion they cause in the fabric of space and time.

A cosmology text should be consulted for detailed analysis of black holes and other space-time anomalies.

Stars

There are many different types of stars. Our own sun is a small, rather average star; there are smaller and larger ones, dimmer and brighter ones. Some stars are hotter than our sun, some not as hot. Some stars are actually almost at absolute zero, having burned out and condensed into extremely dense objects called black dwarfs. Some stars have collapsed so much that they have turned into a solid mass of neutrons, oozing together like warm chocolates into one gigantic neutron. These stars are surrounded by ngaldi formations that warp space considerably from the Euclidean. There might also be many stars that have become black holes, although positive identification of a black hole has not yet been made.

Stars are discrete, identifiable particles, and in fact they look like particles to an earthbound observer. If you didn't know what the stars were, what would you think they were? You would certainly be able to deduce that they were very high in the sky, since the stars look just about the same from any place in, say, the state of Nebraska. But it is not likely that you would imagine them to be as big, bright, and distant as they actually are. Even the brightness of the sun, seen from a few thousands of miles above its "surface," is incomprehensible. Imagine a field as bright as the disk of the sun looks to the naked eye—so bright that retinal damage will occur in just a couple of seconds if you look directly at it—covering fully half of your perceived celestial sphere. Imagine if the ground you stand on were aglow with such brilliance! This is what the sun would look like from a close-up spaceship, if the ship were built to withstand the million-degree heat. You cannot imagine such brightness. It overloads the senses just to gaze at the sun from this planet. Some stars are hundreds or thousands of times brighter than even this.

Stars, like galaxies and the greater cosmos, evolve. They are born from clouds of gas and dust and go through stages, lasting from a few hundred thousand years to billions of years. Strangely, the largest of the stars have the shortest lifespans, often exploding and briefly attaining the luminosity of whole galaxies. The smaller stars are stable, and although they are much less brilliant, their long lifespans are thought to allow for the development and evolution of life on some of their planets if other conditions are right. Astronomy texts describe the most recent theories of stellar evolution. Our own sun has billions of years remaining in its life cycle, so it is thought.

Star Satellites

The next largest type of particle is the planet that either orbits a star or is free-floating in interstellar space. Our own earth is a small planet, about 8,000 miles in diameter. Some planets might be only about 3,000 miles across, for example Mercury or the moon. The difference between a planet and an asteroid is an arbitrary cutoff. Planets orbiting larger planets, such as our moon or Ganymede or Titan (moons of Jupiter and Saturn, respectively), are usually called moons, no matter how big they are. The planet Jupiter is 88,000 miles in diameter, or about 11 times the diameter of the earth. The largest asteroid, Ceres, is almost 500 miles across, although this is along its longest axis.

Asteroids range in size from the large Ceres down to particles the size of dust grains. When an asteroid enters the atmosphere of our planet, we call it a meteor. If it does not completely burn up before it strikes the surface, it is a meteorite. Meteorites are responsible for many ancient, and now eroded, craters on the surface of the earth. The large, relatively recent crater in Arizona is testimony to the power of impact such an object may have. The moons of Mars are small enough to be considered asteroids.

Planets and asteroids contain various substances, including iron and silicates. Another type of small object, the comet, is of still-unknown composition, although it is supposed that they resemble rocky snowballs. These are the objects that develop tails when they come near the sun. We occasionally see comets as spectacular displays in the morning or evening sky.

Space is far from empty. It is literally full of debris. These particles could present a hazard to future space travelers. It is possible that the mishap of the Apollo 13 mission was the result of a collision with a small meteor.

The solar system is in fact so peppered with meteors that on a very dark, clear, moonless night, the counterglow "gegenschein" resulting from reflection of sunlight is sometimes visible. This fuzzy glow appears exactly opposite the sun in the celestial sphere. It occurs for the same reason you see a counterglow from an aircraft on a sunny day. (Look for the shadow of the plane when it is flying at a moderate altitude and you will see this counterglow if the terrain is right.) Particles appear brightest when seen from exactly the same direction as the light source.

Molecules and Atoms

Scientists suspected that matter was made of countless tiny particles long before such particles were actually observed. Molecules, as they are called, are always moving because of kinetic energy. The higher the temperature of an object, liquid, or gas, the faster the molecules will be moving. The molecules collide with each other and the molecules of other substances frequently.

Molecules are too small to be seen with ordinary microscopes. Normally, an electron microscope is needed to observe them. Some molecules, although they are small, are quite long (but thin). Generally, the higher the atomic numbers of the elements comprising a molecule, the larger the molecule will be, and the heavier it will be.

Molecules are in turn made up of atoms, each atom being identifiable as a specific element. There are more than 100 different elements. The element is determined according to its atomic number—the number of protons in the nucleus.

Atoms, like stars, galaxies, planets, and meteors, come in a wide variety of sizes. But, unlike the celestial bodies, there are only certain ways in which atoms can be structured. A given element might occur in various slightly different forms; the details of atomic structure are not of special interest here. As the particle gets smaller, the variability becomes less and less. All atoms are made up of protons, neutrons, and electrons. Only these three, and no others. There are many different kinds of particles that occur free as radiant energy or in other ways; the *photon* is a good example. There are *neutrinos*, which are thought to have no mass at all (although some recent studies indicate they have a very tiny mass). There are *tachyons*, supposedly traveling at speeds greater than the speed of light. There are *quarks*, the constituents of the protons and neutrons that make up every atom.

Atoms are, in their simplest model, something like miniature solar systems, with the electrons orbiting around the nuclei (Fig. 2-2). This was the original concept and has been modified somewhat to account for actual experimental observations. An electron is thought to occupy a specific point in space at a specific instant of time, but their motions are so fast that we can only speak of probabilities for their positions. Electrons seem to be locked into certain orbital "shells" or energy levels within atoms. The rules of behavior of subatomic particles are immensely complicated. The scientist with any interest in theology might ask, "What did God think he was doing when he made things this way?"

Despite the vast number and type of subatomic particles, matter is still almost entirely empty space. If all of the particles were squeezed together so that the space between them

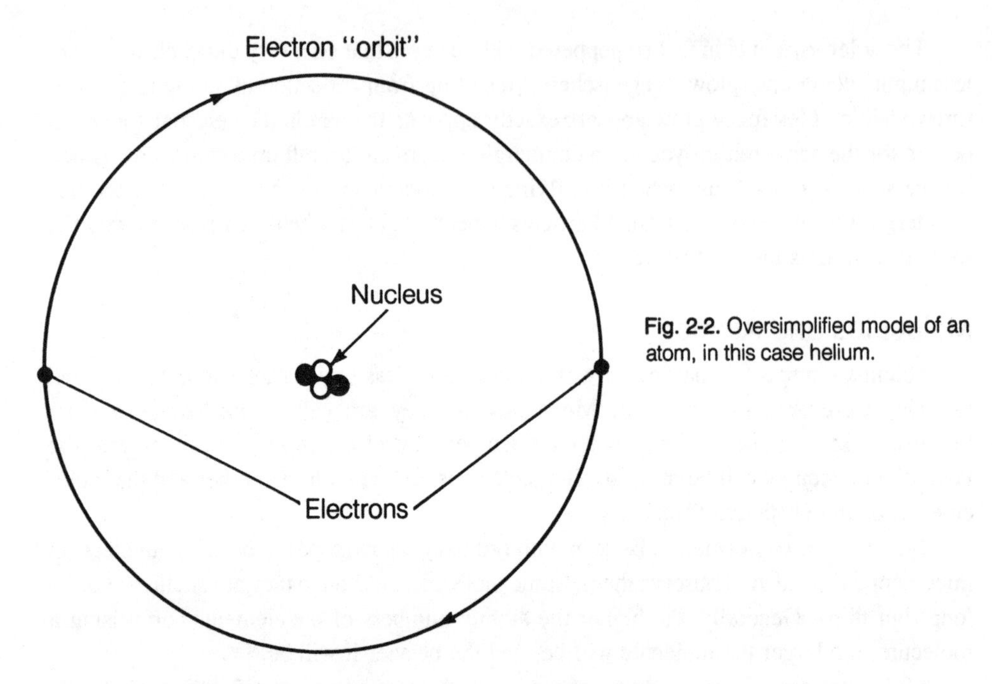

Fig. 2-2. Oversimplified model of an atom, in this case helium.

were gone, most common household objects would be microscopic in size. But they would still mass the same. This is what happens to matter in collapsed stars.

Suppose we could compress a large skyscraper down to the size of a pinhead by removing all the space from between the atoms. The result would be that it would fall through the earth toward the center, because of gravitation. Once the object reached the center of the earth, momentum would carry it on past, part way to the other side of the planet. Then the object would fall back toward the center again, and go past again, and this would continue till the object finally settled at the center of the earth. It would pass through the earth like the proverbial hot knife through butter.

Most of outer space is empty except for countless particles in the form of stars, asteroids, planets, interplanetary and interstellar dust and gas, and an unknown number of black holes and other peculiar objects. Most of the chair you sit in or the bed you lie on is empty space also, with innumerable protons, neutrons, electrons, and the subparticles and sub-subparticles comprising them. In this way, since the particles in space and those in so-called solid objects are in constant motion, outer and inner space are very similar. It seems that the fundamental rule of material things is that they are made up of particles, with mainly empty space accounting for their physical volume.

This might be true of *every* particle, no matter how large or small. It could be as true of a quark as it is of a galaxy.

The Unknown Large and Small

Some philosophers have worshipped the pursuit of knowledge as an end in itself. It has been said that scientists would no longer be happy if all of nature's secrets were revealed. Some have gone so far as to say that the secrets of the universe will never be completely known, that this is a fundamental truth. It would seem that the mathematician Kurt Godel was the most profound in this respect. He proved that there are certain statements whose truth value is not decidable. If this is true in mathematical logic, a universe regarded as antiseptic and pure by the physicist, then how much more true it must be in the actual cosmos, with its complications unending!

The smallest known particles are almost certainly not fundamental particles of the universe. We would be extremely arrogant to assume that we have come to the stopping place now. The science of physics suffered from a dose of exactly this type of overconfidence around the turn of the century from 1899–1900. This was just before Albert Einstein shook the foundations of physics. And physics is not unique in this respect. Mathematics, and set theory in particular, has endured some of the same kinds of shocks. It is the realization that these upsets could occur that keeps things interesting.

Either there is a smallest particle or there is not. This is a sufficiently all-encompassing statement so that it can be regarded as an absolute truth. It appears trivial, perhaps, at first glance, but it is in fact a key to knowledge of particle physics in the future. If we do not find a smallest particle, that doesn't mean there is no such thing; and if we think that we have found it, we cannot be certain because nothing in the physical universe is that certain.

But either there is a smallest particle, or there isn't. Our knowledge does not change the fact.

If we look also to the unknown large, we see the observable universe as the largest "particle." In a certain sense, this is true. We cannot travel outside the three-space surface of the four-sphere on which we live—not by any means that we know of yet. But the non-Euclidean (curved) nature of three-space logically implies the existence of four-space and thus the presence of a universe larger than the one we can see and know today. Such a four-space is necessary for three-space to exhibit curvature just as a globe, with its two-dimensional surface, is a three-dimensional object in its entirety.

What of the four-space in which our three-space-surfaced hypersphere resides? Is it Euclidean? We cannot say since we have no observed data. But it is possible for a space of any number of dimensions to be curved in infinitely many different ways, while it can be Euclidean in only one way. We must not make the error of saying that this means that four-space is "probably" non-Euclidean or even that it is "certainly" so because the "probability" is infinity-to-one. Either four-space is Euclidean or it is not. If not, then the existence of five dimensions, at least, is guaranteed.

This argument could proceed via mathematical induction to cover the set of all integral numbers of dimensions. If a space is n-dimensional (n being an integer) and non-Euclidean, then there must exist $n + 1$ dimensions, *at least*.

It is difficult to directly envision non-Euclidean spaces of three or more dimensions. We, having three-dimensional eyes and looking at photons that are three-dimensional, cannot readily perceive four-dimensional or higher spaces. But we can mathematically describe them in great detail.

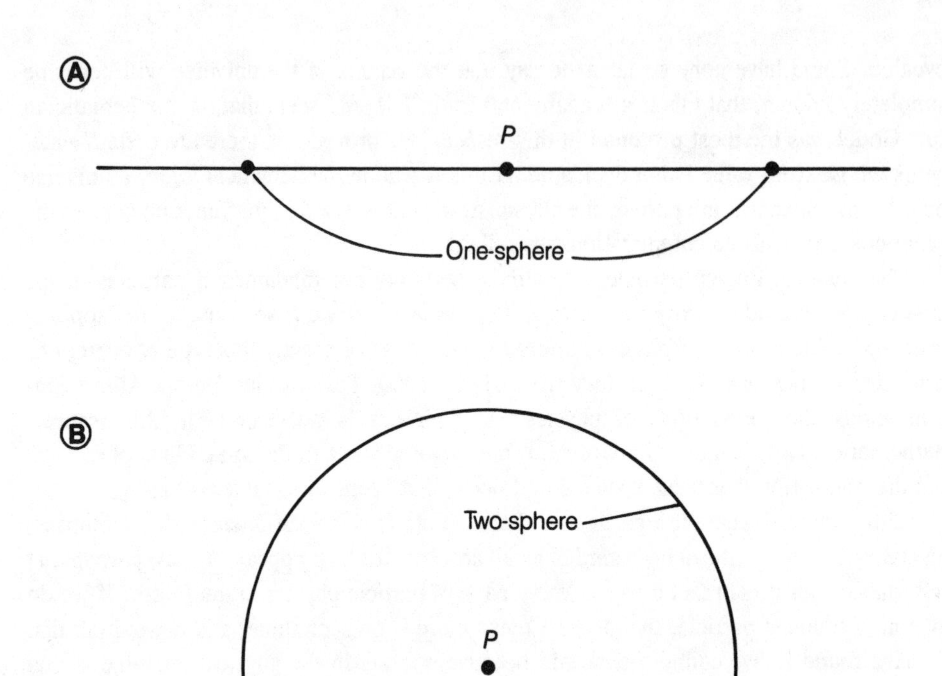

Fig. 2-3. "Spheres" in (A) one dimension and (B) two dimensions.

Consider the case of the *n*-sphere. We define an *n*-sphere as a set of points in *n*-dimensional space, such that each point is equidistant in that *n*-space from a certain center point *P*. The one-sphere is a set of two points (Fig. 2-3A). The two-sphere is a circle (Fig. 2-3B). These can be directly illustrated on a page. The three-sphere is like a basketball or a globe. A four-sphere—well, now, we get into trouble. It is a thing that is shaped, so cosmologists suggest, like our entire universe.* A five-sphere is an object with a four-dimensional surface and is the set of points equidistant from *P* in five dimensions. When we measure the distance that determines the radius of an *n*-sphere, we do so in *n*-dimensional space. This space many not be Euclidean, however, and therefore the path over which we measure the radius might be curved in $n + 1$ dimensions. Consider the example of a two-sphere, or circle, on a three-sphere, or globe (Fig. 2-4). When the two-sphere is very small relative to the size of the three-sphere, the radius lines of the two-sphere are

*Actually, the universe has local irregularities similar to mountains on the surface of the earth, so it is not a perfect four-sphere.

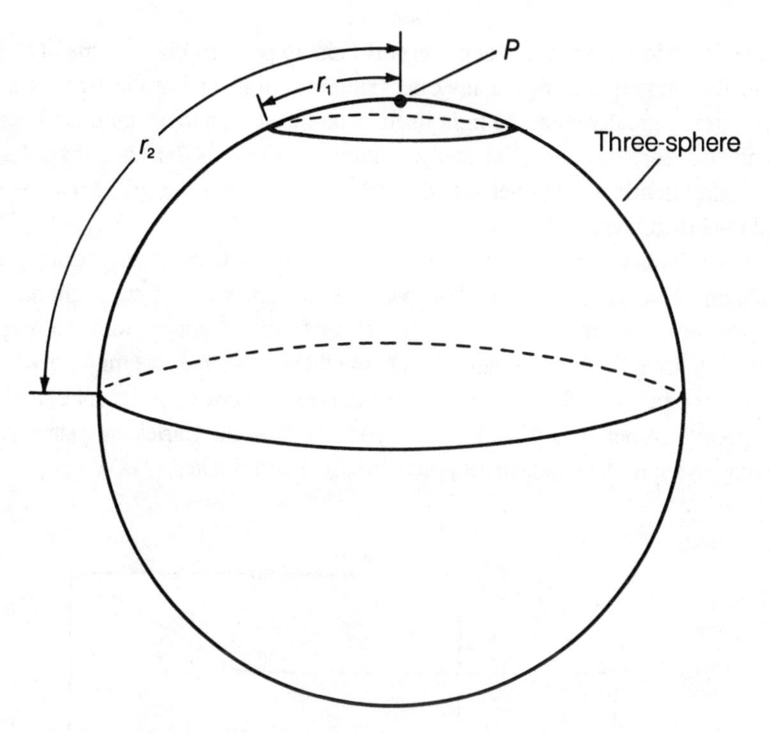

Fig. 2-4. Two-spheres on a three-sphere. The center point is *P*; radii are denoted by r_1 and r_2. The Euclidean value of pi is not accurate here.

almost straight. But it is clear that they are not straight for large circles. We might have the center, *P*, at the North Pole of our planet and the radius, *r*, at about 6,200 miles so that the circle would lie on the equator. It is clear that the formula for the perimeter of the circle, in terms of the radius, does not apply on the surface of the sphere, as it would on a flat plane. It is not equal to the well-known constant relation $C = 2\pi r$, where *C* is the circumference and *r* is the radius. It approaches that relation when the circle becomes very tiny, but as the radius is made larger, the constant, π, normally equal to about 3.14159 on a plane, becomes smaller and smaller. When the radius *r* of the circle is half the circumference of the three-sphere, the value of the "constant" reaches zero. The two-sphere vanishes to a single point.

Similar things happen with three-spheres on the surface of a four-sphere, and, in general, with *n*-spheres on the surfaces of $(n + 1)$-spheres. This brings up an interesting thought. If our universe is really a four-sphere, then all the light from every star must eventually converge on a point exactly opposite on the four-sphere. This would happen the same way the meridians on the globe originate at one pole and converge on the other. Then there should be some points in space where the whole sky is aglow with a brilliance equivalent to that of the surface of the star existing at the counterpoint (point on the opposite side) of the four-sphere.

This will not occur because the half-circumference of the universe is so vast that light has not yet had time to travel halfway around since the star was born. Moreover, the universe is believed to be expanding at a rate sufficient to perhaps prevent this light from ever reaching the counterpoint. Space may be "running away" from the light at a speed as great as, or even greater than, the light itself. Another reason this would not happen, even if the universe were much smaller than we believe it to be, is that the interstellar gas and dust and other debris would attenuate the light so much that the brilliant glow would be reduced almost to invisibility.

Another characteristic of a space of n dimensions is that we need exactly n coordinates to uniquely specify a point in that space. In two dimensions, we might have the Cartesian x and y values, or the polar θ (angle) and r (radial distance from the origin). On a globe we may use latitude and longitude. On the faces of a cube, we might invent a system with x and y coordinates that originate at one corner, or vertex, of the cube, and converge on the opposite corner (Fig. 2-5). In three-space we have the Cartesian system x, y, and z. Or we may use spherical coordinates or cylindrical coordinates.

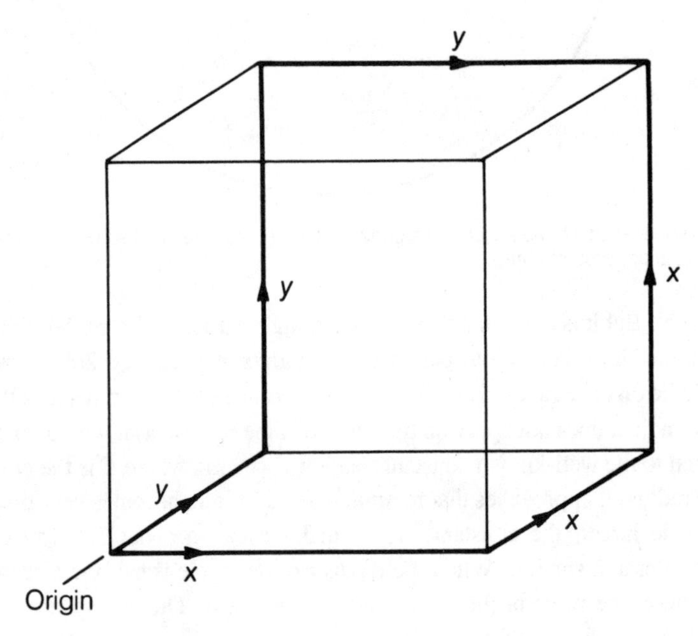

Fig. 2-5. Possible way of establishing two-space coordinates on the surface of a cube. The coordinates are redundant but can be truncated to create a one-to-one correspondence between (x,y) values and points on the surface of the cube.

In n dimensions, we always have the availability of the Cartesian system of coordinates $(x_1,x_2,x_3,\ldots,x_n)$. This system involves axes that are all mutually perpendicular, and all intersect at a common origin point.

It is easy to envision two lines intersecting that are mutually perpendicular. It is somewhat more of a challenge to envision three of them, but consider the point in a room where two walls meet the ceiling. We cannot, however, envision four or more of them directly. And it is impossible even to momentarily conceive of an infinite number of them.

Yet, if each space of n dimensions has even the slightest non-Euclidean characteristic, there follows that there must exist at least $n + 1$ dimensions. Indeed, we can construct a mathematical universe in which there is a dimensionality corresponding to every positive integer, no matter how large. It is but a small step past this to create a space with a countably infinite number of dimensions. Georg Cantor would nod and say something like, "That would be a dimensionality of aleph-nought." *Aleph-nought* or *aleph-null* is the cardinality of the entire set of positive integers.

Consider the problem of naming the coordinates of a point in such an "infinity-space." We could never name them all. There wouldn't be enough time, no matter how long we lived. Even if we were immortal by the standards of our universe, the final contraction (or ultimate entropy) would overtake us and stop our counting sometime, maybe in a few billion years, maybe in a quadrillion years or a googol years. But we would still be naming off the coordinate values of the point in "infinity-space" and would not be finished.

We could get around this by naming the points off faster and faster, doubling our rate of naming every so often, after the fashion of the frog that jumps halfway to the wall in half the time, over and over. But this is not a practical possibility. Our mouths and nerve impulses would have to be moving faster than the speed of light, at least for a brief period of time, and this contradicts the principles of special relativity.

The ultimate "all-that-is" (ATI) may be exactly such a universe: an "infinity-space" such that no point could be uniquely specified in a finite amount of time by mortal beings in a finite-dimensioned subuniverse such as ours.

The Homogeneity Paradox

If there exists a smallest particle of matter, then that particle must be continuous matter. If it is not continuous matter, then it is not the smallest material particle. The term *homogeneous* may be used in place of *continuous*. This differentiates matter and space from mathematical functions and their continuity or lack of it.

There are essentially three possibilities for the fundamental structure of matter:

- There exists a smallest particle, and it is made of homogeneous matter.
- There exists *no* smallest particle. Matter is infinitely divisible.
- Space is quantized, and quanta may be either filled (plus) or not filled (zero).

The case of quantized space was discussed briefly in Chapter 1. We have not yet found a quantum of space or of time. There is a fundamental problem with this idea: How are the quanta moving? (This is discussed shortly.) Let us now proceed with the notion that space is not quantized at all—a good bet, considering the paradox to which such a theory will lead.

Suppose there is a smallest, or fundamental particle. Then it must be homogeneous, or solid, matter. It could vary in density, but it would have to have a finite, measurable average density. The particle would have to have a certain mass and volume, both nonzero and both finite, and this means that its density must also be nonzero and finite.

Could such a particle be squashed? Suppose that gravitation were to reach an intensity so great that all the fundamental particles were packed together. Then if they were squashed, they would become more dense, and they would then not represent the same matter as before. We cannot say for sure what happens when gravitation gets out of control, so this is entirely possible. But the very idea of a particle that could not be squashed without altering the nature of matter is somewhat objectionable. Such a particle could not be cut in half. Even a violent collision between two such particles would not break either of them. This would be true even with the addition of the increase in mass caused by relativistic effects. If we tore such a particle apart, it would no longer be a fundamental particle.

But we can accelerate such particles to almost the speed of light, say 0.999999... 999c, where the string of 9s is as long as the width of the galaxy. The mass would be so great that it seems a collision would surely tear the particles apart—but no. It cannot happen. It is not possible with *any* finite amount of energy, even all of the energy in the entire universe.

That seems incomprehensible.

Besides this problem, there is the notion of homogeneous matter: it is not grainy, not even the slightest little bit. No matter how closely we were to look at a homogeneous particle, it would appear perfectly smooth, although perhaps of nonuniform density within. This, it seems, can only be resolved by considering a fundamental particle to be a point with a force field around it that increases to infinity (that is, grows without bound) near it (Fig. 2-6). Then there is actually a finite mass in a zero volume, and the density is infinite. This is equally beyond our ability to imagine (but to me it makes the most sense).

Suppose that matter is infinitely divisible and that particles get smaller and smaller and there is no end to their numbers. This is a mathematical possibility, but again, is impossible to visualize. No matter how we look at the situation, we have an incomprehensible universe.

The Quantization Paradox

The last possibility is that space itself is quantized into "bits" of space, of nonzero, finite size.

How are such quanta shaped? They might be cubes or some regular polyhedron such as a tetrahedron. But they could not be the most natural shape—spherical—for spheres cannot be packed together to account for all of space.

The main problem with quantized space is of the same nature that was the downfall of the luminiferous-ether theory. That is: how do we determine the motion of a quantum of space? If we can find the motion of such a quantum, then we ought to be able to determine absolute motion in the universe.

Quanta of space would have to be stretchable and compressible if we are to believe Einstein's special and general theories of relativity. Relative motion changes the apparent size and mass of an object but would not change the number of space quanta it would occupy. If a rod is 10^{20} space quanta in length as seen from one point of view, it would

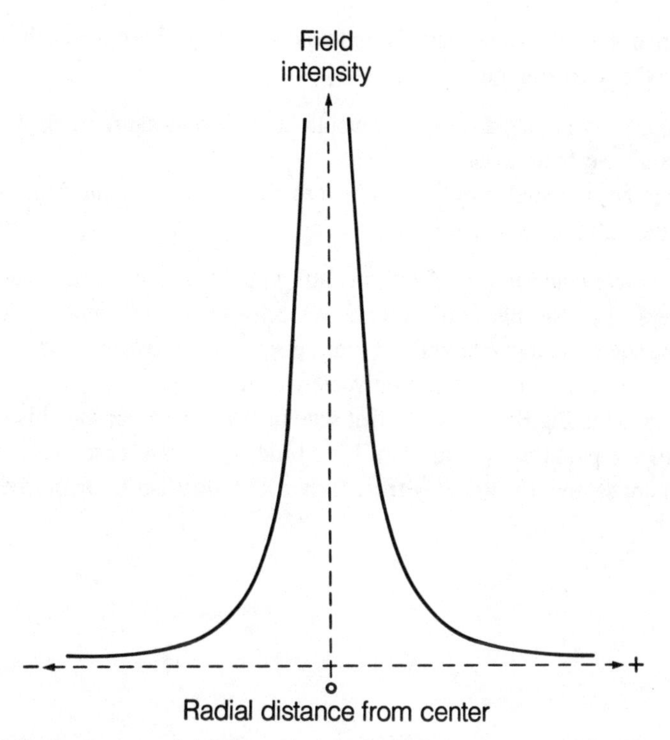

Fig. 2-6. Field intensity versus radial distance for hypothetical elementary particle. The particle itself is point-sized.

have to be this length from all points of view. So space quanta would have to be relative. This is a contradiction with the very notion of a quantum of space.

You might say, "No, the rod would get shorter, and would occupy less quanta as seen from one point of view as compared with another." The rod might appear just 5×10^{19} quanta long from a moving spaceship going at 0.866c but 10^{20} quanta long from the stationary point of view, in accordance with special relativity. This means that there would be a difference in the number of quanta the stationary person would count (if there were time enough to count that high), as compared with the number of quanta the moving person would count. They could rendezvous and disagree on the number of space quanta displaced by the rod. But the rod must have a certain size and a certain number of fundamental particles in it—for space quanta would be, in effect, the smallest material particles.

I suggest that the idea of quantized space be put away, just as Einstein suggested that the idea of luminiferous ether be put away. It's too inelegant.

An Apparently Consistent Idea

The existence of a fundamental particle of nature may well fall into the realm of undecidable things. Kurt Godel proved in 1930, at the age of only 24 years old, the famous Incompleteness Theorem that guarantees that we shall never know all there is to know.

However, if a fundamental particle is found to exist, I have a hunch that it will be found to have the following qualities:

- It would be point-sized, that is, it would have zero measure in all dimensions.
- It would have finite mass.
- It would be surrounded by a force field that increases without limit as the distance from the point approaches zero.

Such a particle could never be split in half. The force field would cause any "collision" to always be a near miss, no matter how much energy was supplied. And these particles could be squashed together without limit, even—dare I say it—down to point size for any amount of matter, even the entire universe we know.

According to the Big Bang Theory, that state of affairs was the way this whole cosmos full of puzzles and paradoxes got its start. It's a little weird, but it fits. This is just a hunch of mine. I'm not saying it's probably true. Either it is, or it isn't. Or maybe it's undecidable.

Chapter 3

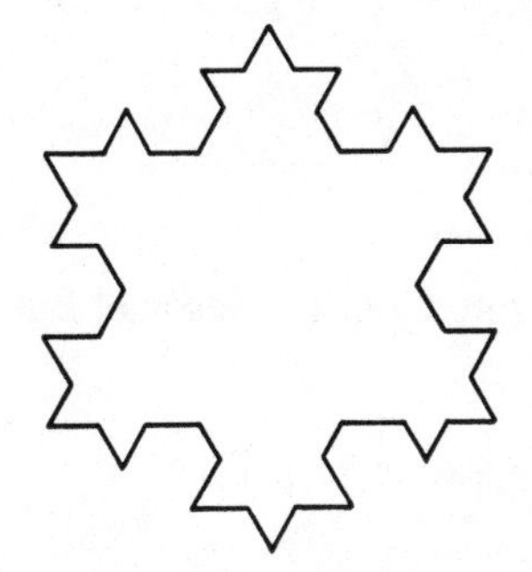

The Number of Numbers

WHEN I WAS A YOUNG CHILD, I WAS UNDER THE IMPRESSION THAT A MILLION WAS THE highest number. Someone said that a hair brush probably had a "a billion" bristles on it, and I objected. That could not be! The largest number was a million. Even though I knew the meaning of "billion"—1000 million or 10^9—I believed there could never be that many of anything. I had closed my mind off at a million. No one ever said, "What if you had a million things and added one more?"The issue was set in concrete. There would be no discussion. It is reminiscent of Medieval attitudes. Some people still think this way, not necessarily about numbers, but they do.

Now I cannot say what the largest number is, because there isn't one, at least not according to what I have been taught. No matter how large a counting number I pick, there can be a number that is one larger.

The Largest Number Nameable

A peculiar paradox occurs when we try to imagine the largest number we can name. The paradox is elegantly brought out in *Infinity and the Mind* by Rudy Rucker.

Suppose we try to figure out what the largest nameable number is. Certain constraints are as follows.

- We must use our native language (English, for those of you who are reading this).
- We must complete the process within our lifespan (say, 80 years).
- We may write to assist our naming if we so desire, but ultimately we must say it in words.

We might begin by announcing: "Let the expression 'ex' mean 'to the exponent of.' Then my number is ten ex ten ex ten ex ten..." and we could keep this up, taking time off only for sleeping and eating, until death. Or we could arrange our lives so that we could get the maximum possible amount of time in rambling: "Ten ex ten ex ten ex..." This has been arranged to minimize the number of syllables. We could perhaps increase the rate by saying "A googol ex a googol ex..." where a googol is 10^{100}. In the first case we would be saying

$$10^{(10^{(10^{(10^{(10^{(10^{(10^{(10\ldots)\ldots)}}}}}}}$$

That, to use a young person's expression, is awesome! But if you consider

$$(10^{100})^{((10^{100})^{((10^{100})^{((10^{100})^{((10^{100})^{((10^{100})\ldots)\ldots)}}}}}$$

—that is super awesome.

The end parenthesis represents death, or the last complete exponent uttered prior to death. (This is morbid, but it illustrates a point.) We can name some awfully big numbers. Perhaps if we were to write such a number as 1 followed by ciphers, the string of zeroes would circumnavigate the known universe several times, or maybe even several googol times.

But someone might live a minute longer, or speak a little faster.

There are certainly natural numbers that cannot be named within the span of a lifetime—aren't there? And if there are, then they represent an *upper bound* for numbers that can be specifically named. This means that such numbers are greater than that number n^*, which is the largest nameable natural number. There must be some specific $n^* + 1$—an actual, unique number—that is the least upper bound for nameable numbers. This fact follows from the theorem of least upper bound, taught in junior high school. This number $n^* + 1$ is the first natural number that is *not* nameable within a human lifespan, or a period of 80 years (we have chosen this arbitrarily). Starting with $n^* + 1$ is a realm of numbers that we cannot know, in a certain sense, because we cannot name them.

Rucker chooses an upper limit of a billion (10^9) words for the length of the diatribe we might use to name a number. He chooses this limit with the idea that it is unlikely that a person would live long enough to utter more than a billion words. Again, this is not a specific number in real life; someone might utter 1,000,003,879 words, and then someone else would come along and utter 1,000,003,880 words. But no one will ever utter a googol words in a lifetime; that we can safely say. Thus, there is a least upper bound for this nebulous variable, as well, and Rucker simply allows it to be 10^9 for convenience.

There is a specific natural number, then $n^* + 1$, that is the smallest natural number not expressible in words. But we have just done so! We have expressed it in a very short sentence. It is true that we have not specifically named it, but we have expressed it. This creates the paradox. We have just described the indescribable. We could generalize this to say that we can describe anything indescribable and that the word itself, "indescribable," is a self-contained contradiction in terms. We can, however, get around this paradox by adding a condition to the previous three:

- We must name the number specifically, to the unit digit, or at least imply that the unit digit is some number of the set {0, 1, 2, 3, 4, 5, 6, 7, 8, 9}.

To *describe* the number is not sufficient. We must specifically name it. We can prove that there exists such a specific number by the theorem of least upper bounds. We cannot, necessarily, state that number. I am not about to begin now; it is much too late. Perhaps if I had begun when I was just learning to speak, and then, may if I had learned to speak fast and my first words had been "Googol ex googol ex googol..."...

All the Natural Numbers

The set of natural numbers, sometimes abbreviated by the capital letter N, is the set {0, 1, 2, 3,...}. Sometimes it is denoted as the positive integers: N = {1, 2, 3,...}. These numbers are sometimes called the *counting* numbers because we can denumerate them easily and directly, or we can get the process started. I recall counting up to 1,000 when I was a child, out loud, one digit at a time, and not missing any. It took a while—long enough to drive anyone who might be listening away from me, I'm sure. In theory I could still be counting. I would be doing so in a mental hospital, but I could be. Maybe I'd be up to about 300,000,000 by now if Rucker's proposition of 10^9 words in a lifetime is reasonable.

But never would I even begin to come close to approaching *infinity*, the accomplishment of having named all of the natural numbers.

Suppose we write random digits, such as listing the digits of the decimal expansion of pi = 3.14159..., and we keep on listing them for the whole of a human lifetime or for the length of time any computer and its clones can spew them out. Suppose we keep doing this until the reams of paper on which the numbers are written completely covers the earth, then fills up the solar system, the galaxy, the local group of galaxies—the whole known four-sphere surface of the universe. Even if this were possible, we would be no closer to an exact listing of the decimal expansion of pi as compared to simply writing the brief start to it, as above. Because of this strange, mind-blowing fact, some people have expressed doubt that we can really say that all of the natural numbers even exist.

In mathematics, we say something is true for a whole set of elements if it is true for each of its members. For the set N = {0, 1, 2, 3,...}, we could ask: does zero exist? Yes. Does the number 1 exist? Yes. If we are given a natural number n in the set N, does $n + 1$ exist? Yes. By mathematical induction, then, every natural number exists. This number is not finite because it cannot be listed in its entirety; but the whole set, as a concept, does indeed exist, even if only because we can think that it does.

It should be noted that mathematical induction is an accepted rigorous method of proof unlike the "inductive reasoning" sometimes used by philosophers, a method that dictates that something is "probably" true but does not guarantee it. Yet the final analysis of this question, "Does N exist in its entirety?" is still a philosophical hurdle.

How many natural numbers are there? If there were a highest one, what would it be like? We can probe this question a little bit.

First, consider the first n elements of the set N. The first one element forms set $\{0\}$; the first two elements form the set $\{0, 1\}$. In general, the first n elements form the set $\{0, 1, 2, 3, \ldots, n - 1\}$. We call the number of elements in a set its *cardinality*. Thus, the set $\{0, 1, 2, 3, \ldots, p\}$ has cardinality $p + 1$ for any natural number p. Likewise, a set of cardinality n, consisting of natural numbers listed consecutively beginning with zero, has a maximum (largest) element $n - 1$. So what?

We can call the cardinality of the entire set of natural numbers infinity, in a sense, because it is not finite. We cannot name this number, if indeed it is a number at all. We cannot even begin to name it, except to call it "infinity." Georg Cantor called it "aleph," the first letter of the Hebrew alphabet, and used the symbol $\aleph$ to denote it. To stick with convention, that is the notation we will use rather than the lemniscate ∞.

With cardinality n in a "counting" set $\{0, 1, 2, 3, \ldots\}$, the last natural number we will name is $n - 1$. This is true for all n; suppose it is true even if we are somehow able to compress our counting process to name an infinite number of numbers in a finite time. (We could halve the interval between naming numbers with every new number. This is not practical, but it is theoretically possible in our minds.) Then if we name $\aleph$ numbers in, say, 32 seconds, the last number would in some sense be $\aleph - 1$.

In fact, if n is any natural number, then $\aleph + n = \aleph - n = \aleph$. This can be proven by considering that you can add or take away any natural number of numbers from the beginning of the sequence 0, 1, 2, 3,.... Beginning with any natural number p so that the sequence is $p, p + 1, p + 2, p + 3, \ldots$, this sequence is infinite and can be paired one-to-one with the elements of the set N of all natural numbers. Therefore, adding the natural numbers less than p makes no difference in the total number of elements. The pairing can be illustrated as in Fig. 3-1. Here we have the specific case for $p = 4$, so the sequence on the bottom begins with the number 4. The pairing with the set $\{0, 1, 2, 3, \ldots\}$ and $\{4, 5, 6, 7, \ldots\}$ is a matter of adding 4 to each natural number n in the original set. The extra four elements 0, 1, 2, and 3 in the upper set do not affect the total cardinality of the set, nor does taking any finite number of elements away have any effect. The theorem thus follows:

THEOREM 1. $\aleph + n = \aleph - n = \aleph$, where n is any natural number.

Likewise, multiplying $\aleph$ by any natural number p, except zero, yields the value $\aleph$. Consider the sets paired off as shown in Fig. 3-2A. In this specific case, $p = 3$. The upper

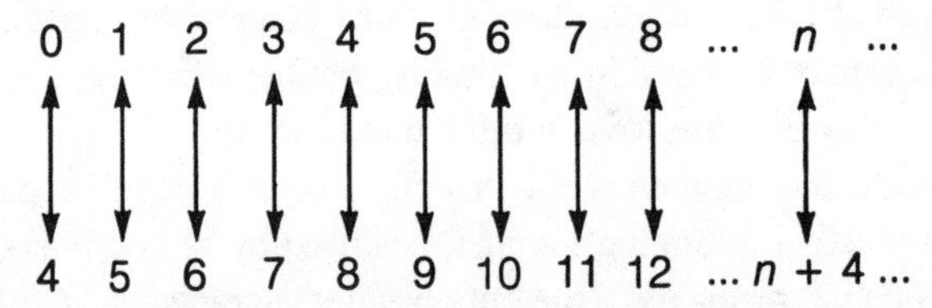

Fig. 3-1. Pairing of the natural numbers with the natural numbers starting with 4. Note that a one-to-one pairing is possible, even though the top set would seem to have four "extra" elements.

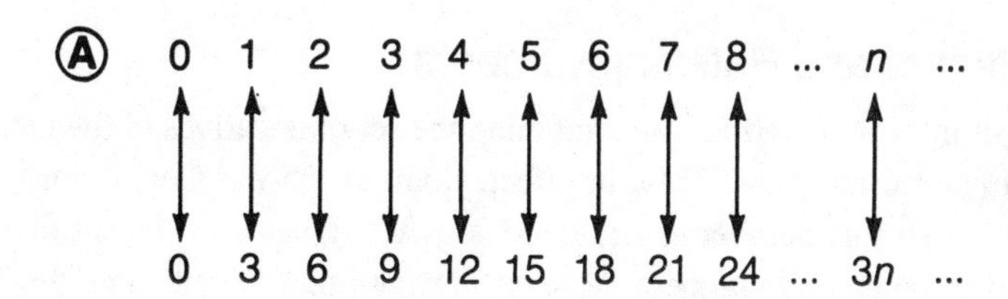

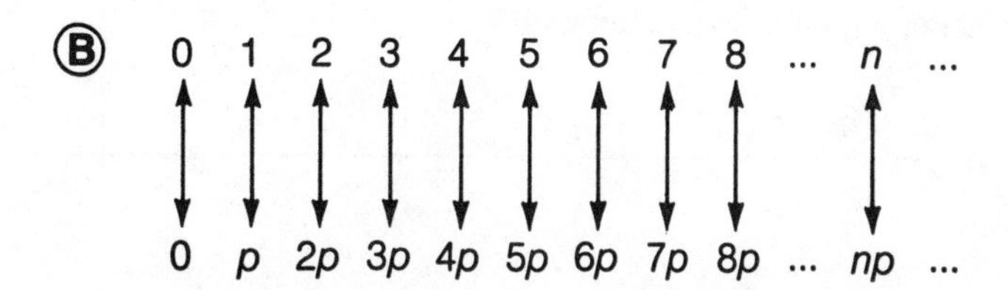

Fig. 3-2. (A) shows one-to-one pairing with the natural numbers and their multiples of 3; (B) shows one-to-one pairing of the natural numbers and their multiples of *p*, where *p* is any natural number except zero.

set, N, is paired off with the lower set, which we might call $3N = \{0, 3, 6, 9, \ldots\}$. We can do this with any multiplier p in the set N, obtaining for the lower set $pN = \{0, p, 2p, 3p, \ldots\}$. The sets are always paired off one-to-one so that for any natural number p, pN has the same cardinality as N. It would seem that N has p times as many elements as pN; this is in fact easy to see by examining the set compared as at Fig. 3-2B. But the total number of elements, and thus the cardinality, is the same for both sets, with the following result.

THEOREM 2. For any natural number $p \neq 0$, $p \aleph = \aleph$.

Although it is more difficult to illustrate, we can generalize this theorem to include any real number r except zero so that $r \aleph = \aleph$. From this, it follows that for any real number r except zero, $\aleph / r = \aleph$. Rigorous proofs of these facts are not given here because that would make this look like a mathematical dissertation or textbook, which it is not intended to be. Proofs of facts concerning this "countable infinity," or aleph, are found in texts on advanced set theory.

Cantor noticed some peculiarities about his infinity. First, it is not a number as such. It is not a member of the set N. For any n in the set N, $n \neq \aleph$. Thus, this value aleph, which is not a number but yet is in a certain sense treatable as a number, had to get a special name. Cantor decided to call it a *transfinite cardinal*.

The transfinite cardinal aleph is different in other respects from the generalized concept of "infinity." It does represent "infinity" in a certain sense—namely, the number of natural numbers—but this might not yield the same result as thinking of "infinity" in other ways, such as the number of real numbers, or the tangent of an angle measuring 90 degrees, or the value 1/0, or any value that, in general, is unchanged when 1 is added to it or subtracted from it. The type, or characteristics, of an infinity, Cantor thought, might well depend on the way in which this quantity was defined and the methodology by which it was derived. This has proven to be so.

All the Integers and Rational Numbers

The set of integers is formed by combining the set of negatives of the natural numbers to the set of natural numbers. This, in effect, doubles the number of nonzero numbers. Since $-0 = 0$, so this number is equal to its own "image" in the set of integers. The integers, unlike the natural numbers, have no starting point. There is no beginning or end point in this set. Figure 3-3A shows the set of natural numbers denoted as points on a line; the set of integers is shown at Fig. 3-3B. We denote the set of integers as $\pm N$ rather than I, to avoid confusion with the Roman numeral I.

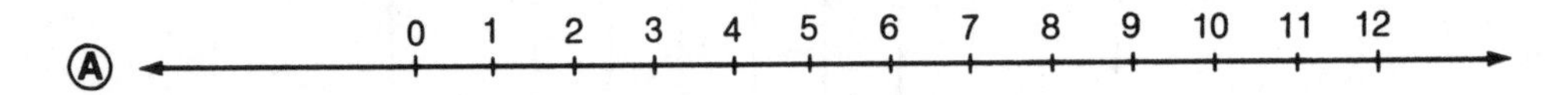

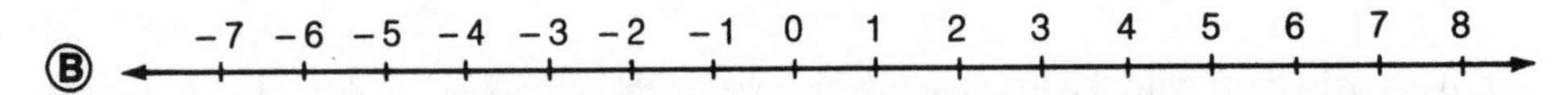

Fig. 3-3. (A) shows the set of natural numbers denoted as discrete points on a line; (B) shows the integers so denoted.

Initial objections to having negative numbers are easily overcome. An argument might proceed: "We cannot have negative numbers. They are meaningless. How could you have negative three apples, for example? It is silly." Yet, there are negative numbers in the temperature scale, and we can express accelerations as negative when decelerating.

How can we create a one-to-one correspondence between the set N and $\pm N$? That is simple. One method is shown in Fig. 3-4. Begin with zero and proceed as 0, 1, −1, 2, −2, 3, −3,.... We don't have to be able to finish the process but only to denumerate it to show how it can be carried on. The cardinality of the set of integers is therefore the same as that of the set of natural numbers: $\aleph$.

A rational number is a number that can be expressed in the form a/b, where a is any integer and b is a natural number and $b \neq 0$. That is, a rational number can be expressed

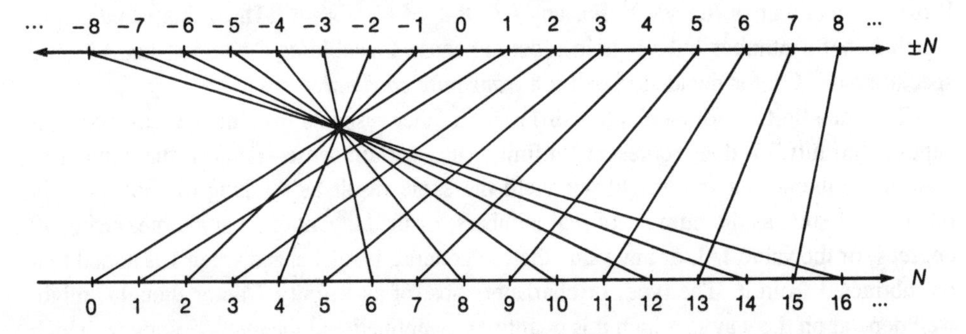

Fig. 3-4. One-to-one pairing of the natural numbers (below) with the set of integers (above). This is just one of infinitely many possible pairings.

in the form of a fraction. All integers are rational numbers with denominators $b = 1$. Some other rationals are integers, also, such as 2/2, 6/−3 and −800/−200.

The set R of all rational numbers is, like the integers and the natural numbers, denumerable. We can actually pair off the rational numbers with the natural numbers. To show how this is done, we can arrange the rationals in an array as shown in Fig. 3-5. The positive rationals go down and to the right; the negative rationals up and to the left. First denumerate the positive rationals by proceeding according to the arrows, skipping over rational values that represent repeats of previous values. For example, skip over 2/4 since that is the same as 1/2, previously encountered; skip over 8/4 because we have already come across the number 2. To denumerate the entire set of rationals, proceed by crossing a positive number and then its negative; for example, you cross −8/3 immediately after 8/3. We thus alternate between two mirror-image paths. This can be carried on without end, establishing a one-to-one pairing between all of the rationals and the numbers 0, 1, 2, 3,....

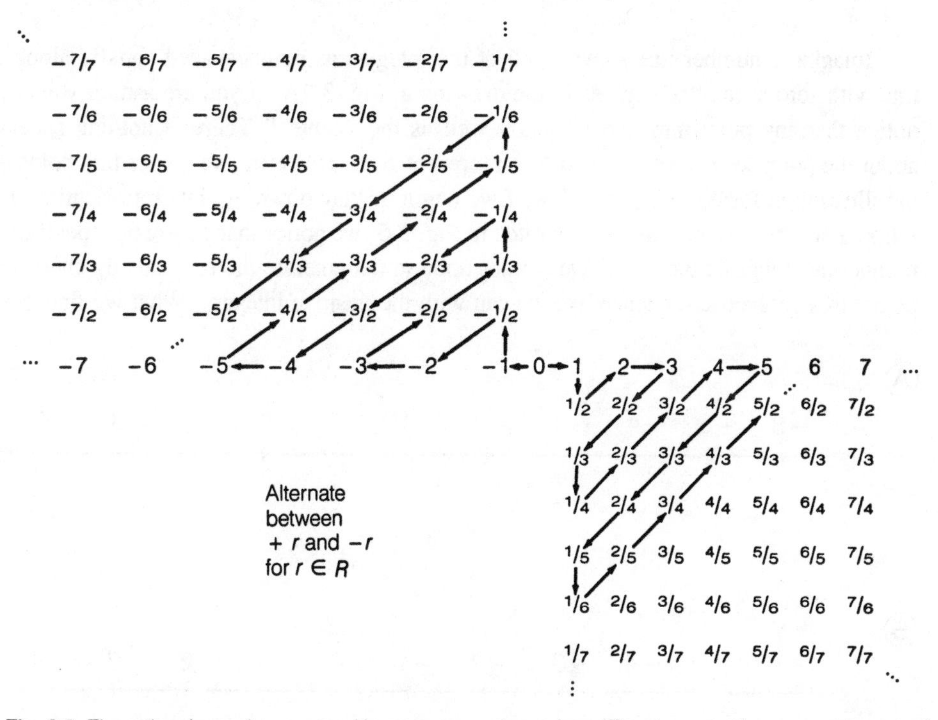

Fig. 3-5. The rationals can be arranged in an array as shown here. The denumeration proceeds according to the arrows, skipping over previous values. For example, we skip 2/2, 3/3, and so on because we have already crossed the number 1. We alternate between all the positive rationals and their negatives in turn, crossing $-r$ immediately after r where r is in R.

This might at first seem impossible. The rational numbers are a sort of "dense" set in that between any two rationals, we can always find another rational according to the formula

$$r_3 = (r_1 + r_2)/2$$

where r_1 and r_2 are the initial two rationals and r_3 is their arithmetic mean (Fig. 3-6). We cannot do this with the set of integers. There is no integer between 203 and 204 or between -89 and -88. How can the set of rational numbers have the exact same cardinality as the set of integers? But it is nevertheless true, despite its being counterintuitive. This kind of thing is a demonstration of the reason why we cannot always simply rely on our intuition to obtain mathematical truths.

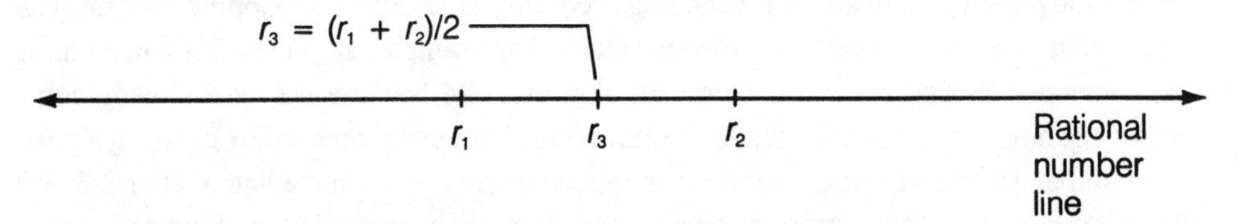

Fig. 3-6. For any two rationals r_1 and r_2, there exists a rational r_3 between them; we simply find the arithmetic mean (average) if $r_1 \neq r_2$.

Imagine a number line showing all of the integers as points spaced equally along a line with zero at the "center," as in the drawing at Fig. 3-7A. If you are astute, you will notice that any point may serve equally well as the "center." There is nothing special about the point we happen to choose to represent zero. However, we center this point in the illustration for convenience. Now if we begin to "lay down" points representing the rational numbers as they are denumerated in Fig. 3-5, we notice that we are not specifying points one alongside the other. We cannot hope to denumerate the rationals by choosing points in a geometric sequence like we can with the positive integers. What we find our-

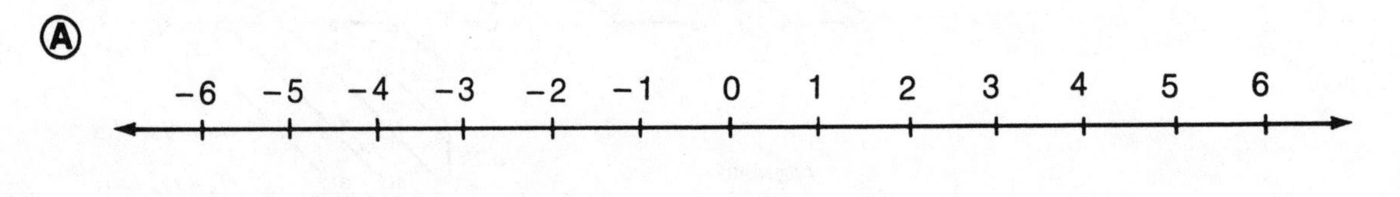

Fig. 3-7. (A) shows the integers denoted as points on a line; (B) shows the first few rationals in the denumeration of Fig. 3-5; and (C) shows the process after a few more steps.

selves doing is "laying down" the points in an ever-finer "dust" (Figs. 3-7B and C). The farther along in the process of denumeration that we go, the finer this point "dust" gets. The whole set of rationals thus corresponds to a "dust" of points so fine that it looks just like a geometric line—an imaginary thing, admittedly, but capable of being visualized in the mind despite its infinite thinness.

As strange as it may seem, we can denumerate the points in a plane by a similar procedure, as long as all the points correspond to *rational* numbers r and s in a Cartesian coordinate system (r,s). That is, if we specify the points in a "rational plane" by means of two rational number lines r and s perpendicular to each other (Fig. 3-8), we will have a set of points that is again denumerable. We can prove the same for a three-space (r,s,t), where the r, s, and t axes contain only rational numbers. This is also true for four-space (r,s,t,u) and in general for any n-space $(r_1,r_2,r_3,\ldots,r_n)$. (Rigorous proof is not given here.)

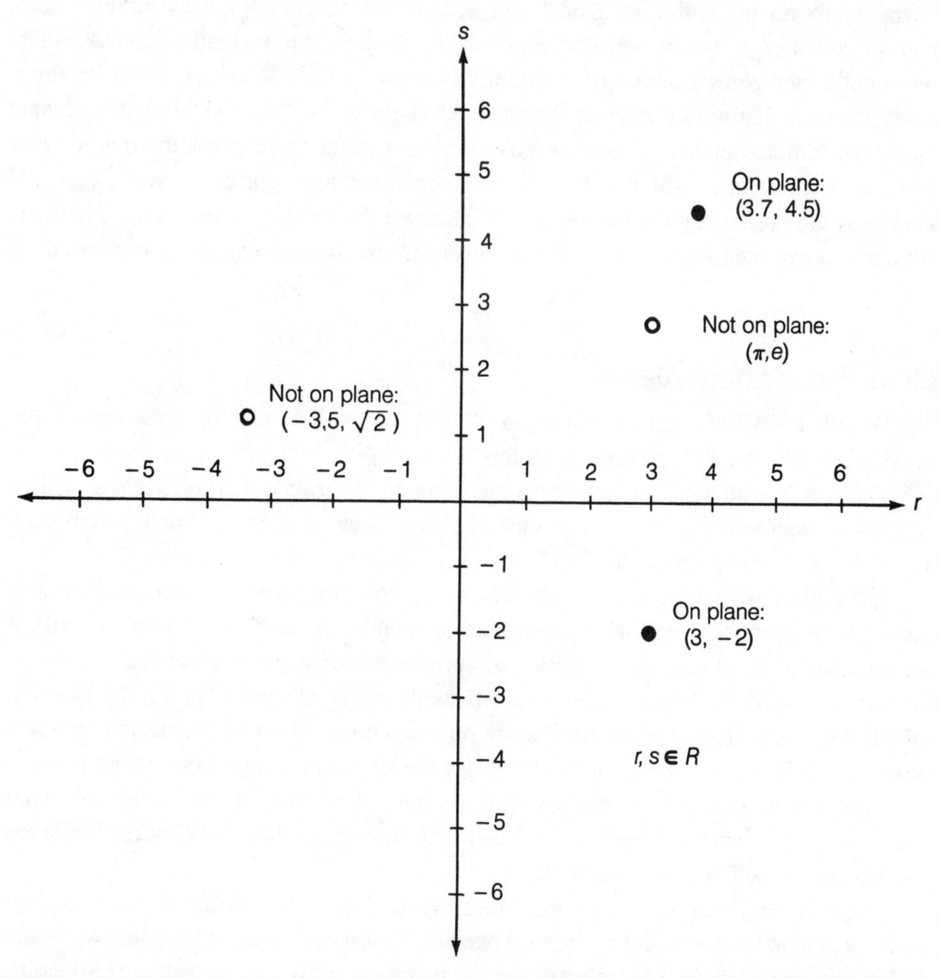

Fig. 3-8. The rationals may be crossed to form a Cartesian plane with axes r and s, both rational number lines. Then certain points (r,s) whose members are both rational will fall on the plane. If either r or s is not rational, the point is not on the plane. Examples are shown with irrational points denoted as open circles.

Rational numbers can always be expressed in a finite number of digits if fractional notation is used. However, if we employ the decimal method of expressing a rational number, some of them will not be able to be written down with a finite number of digits unless we indicate a repeating pattern. For example, the value 34/99 can be written as 0.34343434..., with the understanding that the pattern "34" repeats without end a denumerably infinite number of times. Another way to write this is $0.\overline{34}$.

This raises a philosophical question. Can we actually write out the complete decimal expansion of the rational number 34/99? We can do it if we are willing to accept the implication of the three dots or the *vinculum* (the line over the digits). However, if we require a strict interpretation of actually writing out the decimal expansion of 34/99, we are forced to say that it would require an infinite length of time to do this, and we are not endowed with a supply that great. We can try tricks such as placing the digits in a circle so that there is apparently no end to the string of numbers; however, this is not a truly infinite string, and we cannot say with certainty where we ought to begin any more than we can say where we ought to stop going around such a circle. We can prove that, no matter how many digits we write, we will always encounter the repeating sequence of "343434..."; but we cannot spend the infinite number of seconds, days, or years needed to express the fraction *precisely* in decimal form. Nor would there be room in the known universe, even if we could compress the process by asymptotically speeding up the writing of the digits. There are infinitely many such rational numbers that occur in the form of repeating, nonzero decimals.

Other Facts about Aleph

We might ask the question, "What is $\aleph - \aleph$?" You may already have noticed that there is something a little strange about this.

The way in which we can determine the value of this strange "difference" is to align various sets and see what results we come up with. I was about to say "deliberately misalign" the sets, and perhaps that is a better choice of term.

If we align two sets of natural numbers evenly and then take one member of the second set away every time we take a corresponding member of the first set away, we will, if we can imagine beginning this process at an infinite rate of speed or otherwise compressing our moves infinitely in time, end up eventually taking away the 0 from the first set, followed by the 0 from the second, leaving no elements left over in the second set. This procedure is illustrated in Fig. 3-9A. We begin the snatching-out process infinitely far to the right, and proceed toward the left until we run out of elements in the top set. After taking away the element in the lower set that is directly below the last element in the top set, we have none left in the second set.

But it is clear that we could have "misaligned" the sets, such as is shown in Fig. 3-9B. We might have moved the lower set one notch (number) to the left, so that we would be left with one element after taking away all members of the top set and corresponding members of the lower set. We would in this case be left with the single element 0. Thus we may say that in the case of Fig. 3-9A, $\aleph - \aleph = 0$, whereas in the case of Fig. 3-9B,

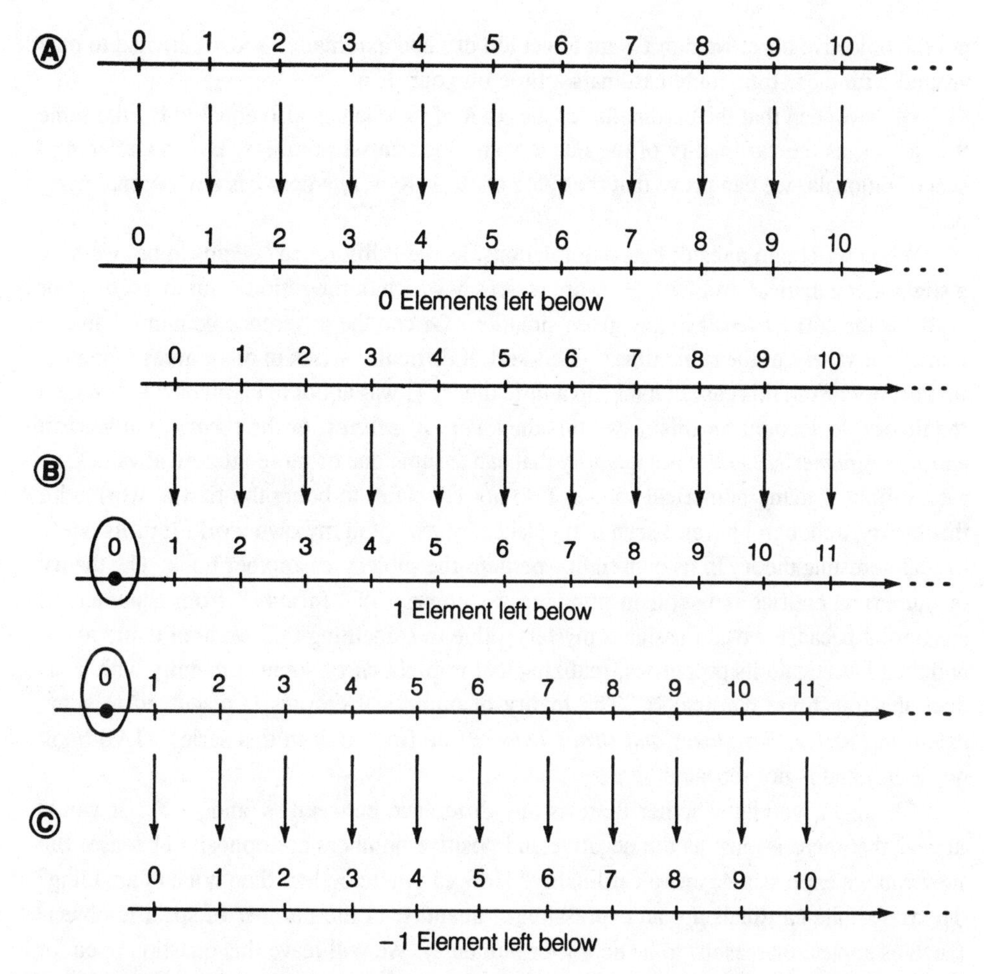

Fig. 3-9. At (A), elements of the top set are taken out first, followed by corresponding elements of the bottom set, leaving no elements (zero elements). At (B), the sets are displaced so that one element (the number zero) remains in the bottom set after all the top elements are stricken out. At (C), the sets are displaced so that we have one element left in the top set with no corresponding element to strike out in the lower set. This may be thought of as having −1 element left in the bottom set.

$\aleph - \aleph = 1$. We can move the sets askew by two, three, or any natural number of elements, and thus get the result that $\aleph - \aleph = n$, where n is any natural number.

We cannot readily imagine doing things the inside-out way, leaving a negative number of elements in the lower set, although we can define this as having elements left over in the top set after we have exhausted them in the lower set. The "set misalignment" that would lead to a result of $\aleph - \aleph = -1$ is shown in Fig. 3-9C. We can, as with the positive misalignment, move the sets askew so as to obtain any negative integer as the resultant. We now have the result that $\aleph - \aleph = q$, where q may be any integer whatsoever.

We have already established that $2\aleph = \aleph$; it is therefore easy to assert that $\aleph - \aleph = \aleph$, and it's not an especially difficult exercise to discover that $\aleph - \aleph = -\aleph$ as well. This begins with recognizing that $2x = x + x$ for any x. Again, complete rigor is not the objective here; we don't want to get too dry and axiomatic. You are invited to play around with these transfinite cardinals a little on your own.

We have seen that the cardinality of the set R of rationals is also equal to $\aleph$, the same $\aleph$, in fact, as the cardinality of the integers and the natural numbers. By "misaligning" sets of rationals, we can prove that the value of $\aleph - \aleph = r$, where r is any rational number.

When we obtain a result that is ambiguous, as the "difference" seems to be, we have a slight mathematical problem. How are we to say which of the rational numbers, or $\aleph$ or $-\aleph$, is the correct result in any given situation? Or can the difference be more than one numerical value "at the same time?" This sort of difficulty arises in other areas of mathematics, not just in this case. I made up a little theory (I was about to claim having invented the theory) to account for this. I call this the *theory of entities*, or the *theory of numerical entities*. A *numerical entity* is a quantity that can assume one or more numerical values, or even infinitely many numerical values. I do not lay claim to being the first to work with this theory, although I haven't seen it anyplace else except in my own work. It turns out to be an interesting theory in its own right—perhaps the subject for another book. The theory of numerical entities is useful in pursuing the concept of "infinity" from a numerical standpoint because we can assign a mystery value to something that we aren't sure about and then investigate its properties, realizing that it might carry some ambiguity with it, as does the transfinite cardinal $\aleph$. The theory of numerical entities is discussed to some extent in *Puzzles, Paradoxes and Brain Teasers* (the first book in this series—TAB book no. 2895) and is not repeated here.

We might wonder whether there is any difference between $\aleph$ and $-\aleph$. It would appear that there is, just as the negative and positive numbers are opposite in sense. But how can we express a negative cardinality? How can there be less than none of anything? The transfinite cardinals are an expression of quantity, of the number of specific objects (such as apples, or reasons to be happy, or numbers). We will leave this question open for the time being.

We can evaluate the product $\aleph \aleph$ by considering it to be the number of points on a Cartesian plane where the axes contain integers or rational numbers. If we consider axes of integers, it is easy to see that the square of the transfinite cardinal $\aleph$ is denumerable and is therefore equal to $\aleph$. Thus it follows that $\aleph^n = \aleph$, for any positive integer n.

We may now ask the question: is there anything that will result in a value that is not finite and not equal to $\aleph$? If so, then we can say that this result is a different "infinity" than the aleph, represented by some other set than the set of all natural numbers, or the set of all integers, or the set of all rational numbers.

It turns out that there are indeed "infinities" that are greater than $\aleph$. We are therefore forced to accept something, again, that is counterintuitive, for it would seem impossible to have something greater than infinity. Just as I could not accept that there could be anything greater than a million in my childish closed-mindedness, so most find it difficult to accept that there could be any quantity truly greater than infinity. A better way of saying this is that aleph, now renamed aleph-null or aleph-nought and written $\aleph_0$, is not the largest transfinite cardinal that can exist.

The ancients thought that everything could be described in terms of numbers. It turns out, as Cantor first found, that a better definition for mathematical objects can be formulated if we consider such objects to be sets. Every real number can be defined as a certain set, based on the rationals, in turn defined as sets of integers or natural numbers, themselves defined as sets of sets. So we are not actually speaking of "infinity" when we speak of $\aleph_0$; rather, we are speaking of the number of objects in a certain defined set—a set that turns out not to be the largest possible set.

How can we prove that there exists a set of numbers larger than the set of all rationals? We must first understand the difference between a *denumerably infinite* set and a *nondenumerably infinite* set. Aleph-null is a denumerable infinity because we can denumerate (express in counting terms) the natural numbers, the integers, or the rational numbers. We have already seen ways in which this denumeration can be carried out.

Irrational Numbers and Nondenumerable Sets

It can be proven that $\sqrt{2}$ is not rational. That is, it cannot be expressed as a fraction a/b, where a is an integer and b is a positive integer.

But we can approximate the value of $\sqrt{2}$ as a decimal. If I take my trusty Japanese light-powered calculator and punch up the buttons needed to get a display of $\sqrt{2}$, I obtain the readout 1.414213562. This is accurate to nine digits (I don't know if the last digit is rounded or not). I'll try and find out if this calculator will delve beyond its own display by subtracting 1 from the display and then multiplying by 10. I will get 4.1421...what? I get 4.142135623. Aha! But this still doesn't tell me if I have a calculator that rounds off or one that merely truncates. When I try again, subtracting 4 and multiplying by 10, there aren't any more hidden digits. It doesn't matter enough to me, that I feel like testing until I find out whether the calculator rounds up or truncates. Well, I know that there are many more digits that the calculator will not show. In fact, even with the most powerful computer in existence, I could never hope to obtain the exact decimal readout for $\sqrt{2}$. Why? Because the string of digits to the right of the decimal point will never end and will never have a repeating pattern. The digits will go on and on forever in a way that defies "prediction." The digits in fact seem to be random, although there is a philosophical debate that can go on regarding whether or not they are truly random. (There may be no such thing as random numbers, according to one argument.)

There are a lot of numbers like this. In fact, since the decimal expansions are infinitely long—they never end—there can exist an infinite number of different expansions. We might denote an expansion as

$$x = 0.a_1a_2a_3a_4a_5...$$

where a_n represents a digit from the set $\{0, 1, 2, 3, 4, 5, 6, 7, 8, 9\}$ and with the condition that no repeating pattern can be found that would allow the number to be expressed as a fraction and thus be rational.

There are some numbers that are irrational yet do have a recognizable pattern of sorts. One example is 0.0100100010000100000 1..., where the number of zeroes constantly increases by one each time before we get the digit 1. However, this is not a repeating decimal of the rational type and therefore it qualifies as an irrational number, even though we can predict which digit will come next after any given number of digits. Likewise, we may have the irrational number 0.1234567891011121314151617..., consisting of all the natural numbers pasted together one after the other. We can again predict what will follow—usually, although I'm sure you can invent a trick example to show that we might not be able to predict the next digit—but we cannot express this decimal as a fraction of two integers.

But can we denumerate the set of irrationals $x = 0.a_1a_2a_3a_4a_5...$? In order to prove that we can, we must demonstrate a way. No one was able to do this, and they gave up trying when it was shown that there *cannot* be a way. This was done via the technique of *reductio ad absurdum*: literally meaning "reducing to a contradiction" or, if you like, "deducing that it's pure baloney."

Suppose it were possible to denumerate all of the irrationals between 0 and 1. According to the infinite-decimal-expansion definition, these numbers take the form $x_1 = 0.a_{11}a_{12}a_{13}...$, where the a_{1n} are digits from 0 through 9 inclusive, $x_2 = 0.a_{21}a_{22}a_{23}...$, where the a_{2n} are digits from 0 through 9 inclusive; in general, $x_m = 0.a_{m1}a_{m2}a_{m3}...$, where the a_{mn} are digits from 0 through 9 inclusive. It should be possible, if the list is aleph-null rows long, to include every irrational between 0 and 1.

We could construct an irrational number $y = 0.b_{11}b_{22}b_{33}...$, where $b_{11} \neq a_{11}$, $b_{22} \neq a_{22}$, $b_{33} \neq a_{33}$, and in general, $b_{pp} \neq a_{pp}$ for all natural numbers p. Thus, we know that the decimal $0.b_{11}b_{22}b_{33}...$, which lies between 0 and 1 as can be plainly seen, is not equal to any of the numbers in the list. For it differs by at least one digit from all of the numbers listed. We deliberately made this new number that way. We could make an infinite number of such numbers not on the list, in fact, since there are nine choices for each digit that would differ from the digit in the appropriate number on the list, and there are infinitely many numbers on the list already.

This is a contradiction, then, because by assuming that we can list all the irrationals between 0 and 1, we can easily construct numbers not on this infinite list. It might at first seem like quite an accomplishment to be able to show that an infinite list might not include certain items; however, it proceeds nicely by mathematical induction. The list and the method of constructing the new number are shown in Fig. 3-10.

No matter how we try to list the irrationals between 0 and 1 or between any two consecutive integers, we will come up against this problem. The irrationals in any given defined interval, no matter how small this interval, cannot be listed. We can always construct numbers not on the list, using the exact same method described above and illustrated in Fig. 3-10. We conclude that the set of irrational numbers is a nondenumerable set—that there are somehow more irrationals than there are elements in any denumerable set. The infinity of the irrational numbers is greater, in this sense, than the infinity of the natural

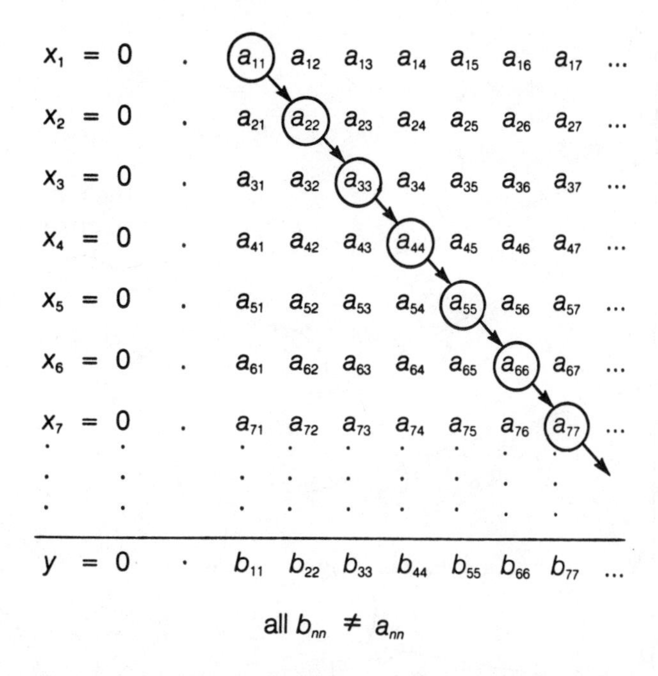

Fig. 3-10. Listing all the irrationals between 0 and 1 is impossible; we can always construct another num ber that is not on the list. The meanings of the various notations are discussed in the text.

numbers, integers, or rationals. Cantor, recognizing this, called the infinity of the irrationals by the name "aleph-one" and wrote it as $\aleph_1$.

The Continuum Hypothesis

We often see the real numbers illustrated as lines, such as the axes on a coordinate system. The real numbers include all rationals and irrationals, in fact, all the numbers of the form $q.a_1a_2a_3...$, where q is an integer and a_n are digits from 0 through 9 inclusive. We say that there exist infinitely many points on a line, and, in fact, in any finite interval of a line (line segment). But how many points are there on a line or a line segment? We can say that for each real number, we can find a point on a line by means of some mapping function. This does not seem to be a problem (Fig. 3-11). There are certainly enough points on the line. We can always squeeze a new point in between two existing ones (as long as the two existing ones aren't identical). But what about the reverse mapping—points on a line to the set of real numbers? There might be too many points on the line. There may be some points that cannot find a place in the set of real numbers. How are we to know whether this is the case or not?

Georg Cantor believed that the *number of the continuum*, or the number of points on a geometric line, is aleph-one. But it remains only a hypothesis, called the *continuum*

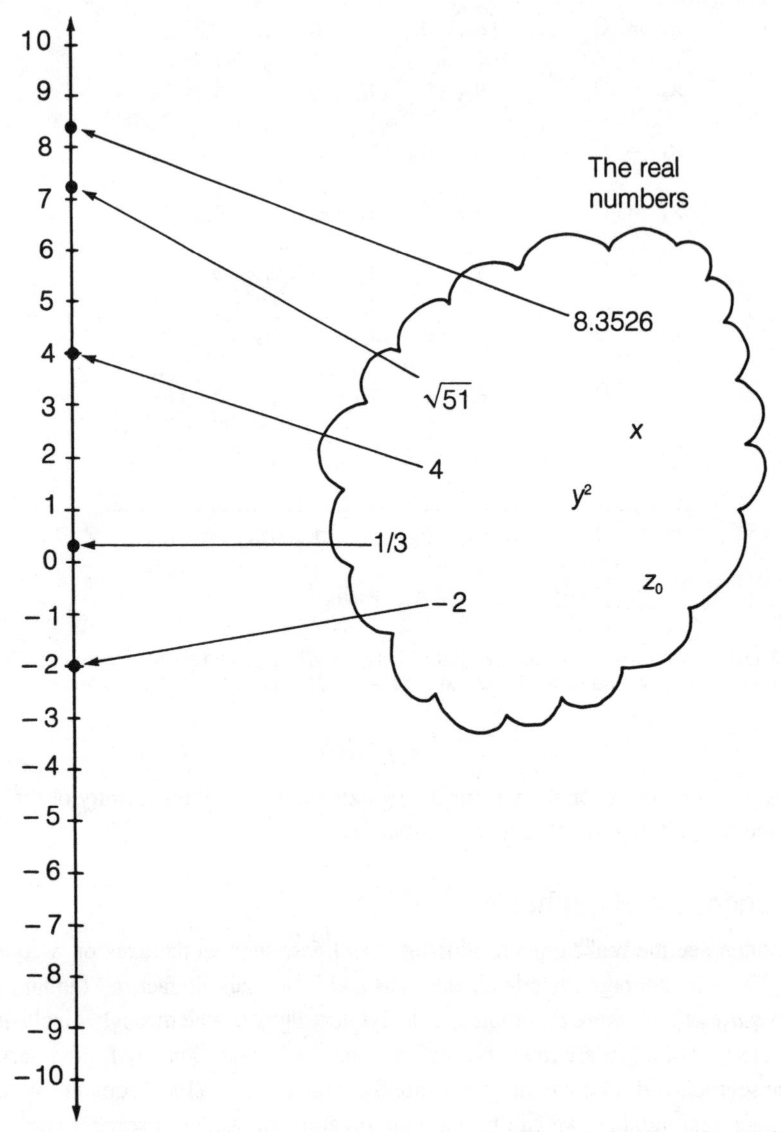

Fig. 3-11. The real numbers can be mapped onto points on a line; this is the familiar concept of the number line. The reverse mapping is a more confusing issue.

hypothesis. We simply do not know for certain whether space can be broken down into any more than aleph-one components.

Do transfinite cardinals greater than $\aleph_1$ even exist? Yes. It has been proven that, for any transfinite cardinal *C*, the value 2^C is greater than C. This opens the way to an infinitely long hierarchy of infinities; we might have $\aleph_n$ for any natural number *n*, and $\aleph_{n+1}$ will always be greater than $\aleph_n$! And we need not stop here. There might be transfinite cardinals such as aleph-aleph-null, aleph-aleph-one, and so on. These would be denoted as $\aleph_{\aleph_0}$ and $\aleph_{\aleph_1}$. We could in fact continue until we have

$$C = \aleph_{\aleph_{\aleph_{\cdots}}}$$

where the string of alephs goes on $\aleph_0$ times before we encounter a subscript!

This gets rather silly after a while; the game can be played until we run out of ingenuity for notation. Ultimately, we encounter the concept of the inconceivable: an absolute infinity, sometimes denoted by the symbol for omega, or Ω, the last letter of the Greek alphabet, symbolizing the end of all things.

This absolute infinity is greater than anything we can conceive and is perhaps the idea that we originally dream of when we think of "infinity."

We have come full-circle. After all of this exploration of transfinite cardinals, their behavior, and the fact that there exists a hierarchy of such quantities, we end up with the same lack of grasp for the ultimate infinity that I had when I was convinced that "a million" was as large as a number could get.

No matter what notation we invent for the hierarchy of transfinite cardinals, there is a limit to the size of such a quantity that can be expressed in a human lifetime. The syllables we utter are, in fact, not only denumerable, but in practice, finite.

We believe infinity exists because we can conceive of it, but can we really, other than to say things about it such as "The ultimate infinity is bigger than anything we can conceive of"? Does the unimaginable exist just because we say that it can?

A little child might say, "Infinity is really big. Bigger than big. Bigger than bigger than big. Bigger than bigger than bigger than..." And, while stretching his arms out as far as they will span, the child runs out of breath, only to draw in another breath, deeper than the last, with which to continue his dissertation.

All of the sophisticated mathematicians actually can do no more than that.

Chapter 4

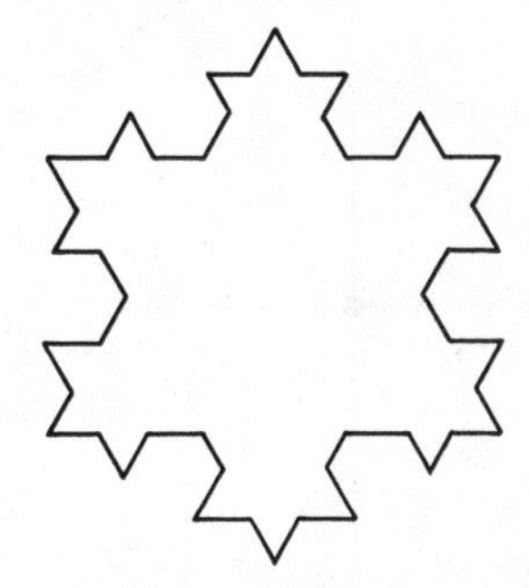

Arithmetic with Infinity

SHORTLY AFTER BEGINNING THE OUTLINE FOR THIS CHAPTER, IT SEEMED TO ME THAT A whole series of books could be written about the various transfinite cardinal numbers. There are so many of them that the process of describing them is sooner or later likely to drive any student to throw his hands in the air and scream. I developed this intuition as I read about transfinite cardinals in Rudy Rucker's book, *Infinity and the Mind.* There exists an enormous variety of transfinite cardinals with a tremendous variation of properties, and the adjectives I am using to describe them seem as firecrackers next to hydrogen bombs for adequacy of effect. I can only recommend that you obtain and read *Infinity and the Mind* to start to get a grasp of these things; that level of sophistication is far beyond anything I wish to attempt here.

The scream might be one of consternation, awe, exultation, and perhaps even disgust or abhorrence: how could God dare to make a universe so utterly complicated? But there is comfort in knowing that we might even attain the infinite and have no fear of running out of things to keep looking for. So if a scream can indicate relief, there ought to be some of that apparent, too.

Division by Zero

The whole business of "division by zero" will be dismissed immediately by some as *not defined* and therefore not worthy of any discussion at all. But it is natural to wonder why such a quantity as 1/0 remains undefined. Geometrically it can be thought of as similar to the value of tan 90 (the tangent of 90 degrees). In a unit-circle model for trigonometric functions, the sine is depicted as the value y on the unit circle and the cosine as the value x; the tangent is equal to y/x. This is shown in Fig. 4-1A. When the measure of the angle θ is 90 degrees, $y = 1$ and $x = 0$. Then $y/x = 1/0$, and the value $1/0 = \tan 90$ by association.

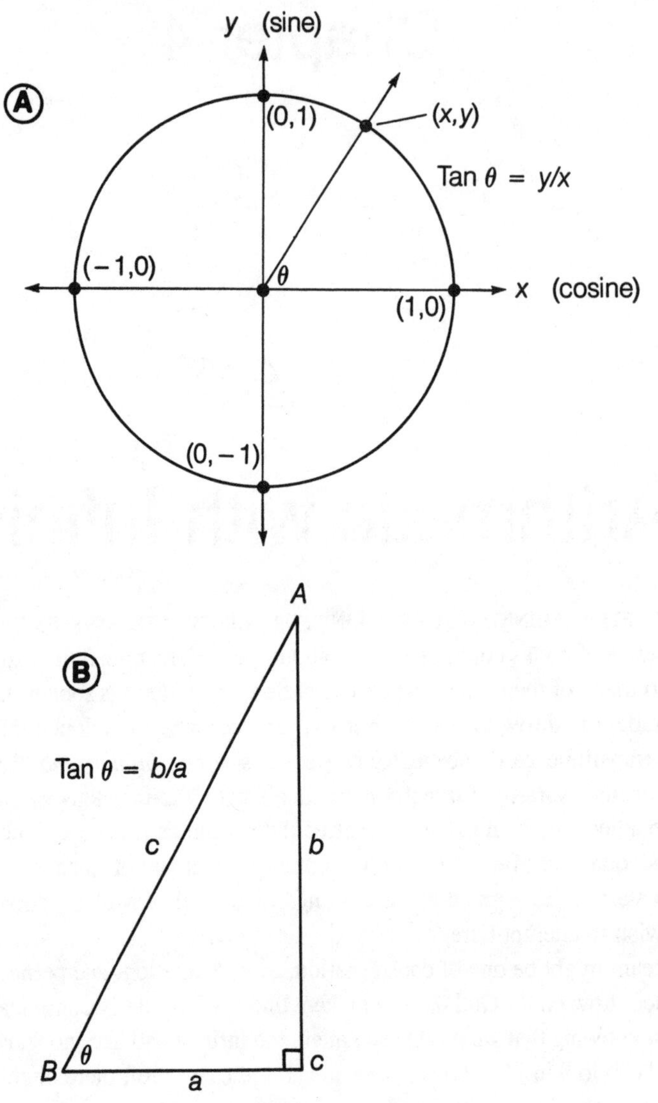

Fig. 4-1. (A) shows a unit-circle model representation of the tangent function and at (B), a right-triangle representation.

But how can we say any more about tan 90 than about 1/0?

There is another way to look at the trigonometric functions, and that is the right-triangle model. In this model, the tangent is the ratio of the opposite side (or the height of the triangle) to the base. That is, it is the ratio of the lengths of these sides, as shown in Fig. 4-1B for an angle of about 60 degrees. What happens as the angle θ is increased towards 90 degrees? The height of the triangle increases without limit. Or, the length of the base must be reduced toward zero. Suppose we allow the height to be equal to one unit and reduce the base to almost zero so that the angle θ is almost 90 degrees. The value tan θ will have a very large size. There is no limit to how large it can get: a million, a billion (10^9), a googol (10^{100}), or even larger. We might want to think that if $\theta = 90$, the value tan θ is equal to "infinity." Then we could say by association that $1/0 = \infty$.

We can also increase the height of the triangle, supposing that the length of the base is one unit, and the angle θ will increase toward 90. The value tan θ is then simply the value of the height of the triangle in base units. This clearly increases without limit to a value of countable infinity, or aleph-null ($\aleph_0$) units. Thus, we might want to say that tan $90 = 1/0 = \aleph_0$.

In *Puzzles, Paradoxes and Brain Teasers* (the first book in this series), I introduce the idea of *contingent equality*. The reason for such a concept becomes clear when we realize that we can perhaps call 1/0 the same thing as aleph-null, but one might not really be identical with the other. It is possible for two entities to share certain properties, which makes it tempting to think that they are identical, but they might actually have other properties that differ greatly. We haven't really even defined what it means to divide by zero, for example, and until we do that, we can't say such things as $1/0 = \aleph_0$. We might ask, "How many zeroes is it necessary to add to one another in order to get one as the result?" The answer is not any finite number of zeroes. For any finite number of zeroes, $0 + 0 + 0 + ... + 0 = 0$. Thus $n0 = 0$, no matter how large the counting number n is. But if we could add up $\aleph_0$ zeroes, we might get something else besides zero. If we could get one, then we could add another $\aleph_0$ zeroes to this and get 2, and we would be able to say that $2/0 = \aleph_0$ as well as we could say that $1/0 = \aleph_0$. This would not imply that $1 = 2$. We could not use conventional algebraic manipulation in every sense that we can use it with real-number denominators. In fact, we could say that $R^+/0 = \aleph_0$, where R^+ denotes any real number or any combination of real-number values "simultaneously." This is an example of a numerical entity equation, where R^+ and $\aleph_0$ behave in some ways like regular numbers and in other ways quite differently from numbers as we are accustomed to them.

Another way to define the quantity 1/0, or $x/0$ for any real number x, is to ask, "What is the ratio of x to 0?" An especially interesting case comes up when we consider the case where $x = 0$, thus attempting to evaluate the ratio 0/0. We might write the equation $0/0 = x$, and then try to "plug in" different values of x after rearranging the equation by cross multiplying $0/0 = x/1$ to get $0 = 0x$. It is immediately clear that any real number r will satisfy the equation. This gives rise to the notion of the entity R, representing all real numerical values simultaneously. This is such an all-inclusive numerical entity that it appears to be trivial and more than a little bit silly to concern ourselves with the value 0/0. But it is at least definable in this manner.

Add One and Get No Change

Another method of approaching the problem of "infinity" as a number or as a numerical entity is if we say that we can't change infinity by adding one to it. Infinity is so large that adding one doesn't make any difference at all.

Even with natural numbers, the larger a number n, the smaller the percentage difference that results from adding one. If $n = 1$, then $n + 1 = 2$, and this is a 100-percent increase. If $n = 10$, then $n + 1 = 11$, and this is a 10-percent increase. If $n = 100$, then n

$+ 1 = 101$, a change of just one percent. And so on it goes. But for any n, it is always true that there is some finite percentage increase when we add one; and always, $n + 1 \neq n$. Of infinity, such is not the case. Not only is there a zero-percent change if we add one, but the actual value is not affected either.

Consider the entity Q of all integer-number values. Does this value change if we add one? Interestingly, the answer is no. Think of

$$Q = (\ldots, -3, -2, -1, 0, 1, 2, 3, \ldots)$$

Adding one to this entity is the equivalent of adding one to each member of the entity. As a result, the entity remains exactly the same. We might depict it as

$$Q + 1 = (\ldots, -2, -1, 0, 1, 2, 3, 4, \ldots)$$

but this does not indicate any real difference and only serves perhaps to illustrate that what we have done really doesn't make any difference.

The entity consisting of all real numbers does not change either if we add one to it; nor does the entity consisting of all rational numbers. You might want to demonstrate this to yourself. How about the entity consisting of irrational numbers? This entity would not change either if we add one to it. The demonstration of this is a little trickier than it is for the previous three cases. I will leave that up to you, too, if you want to try the demonstration.

The entities mentioned above are not actually the same thing as we think of when we think of infinite quantities. It is true that there exist an infinite number of numerical values in each of these numerical entities, denumerably infinite in the case of the integers and also in the case of the rationals, and nondenumerably infinite in the case of the reals and the irrationals. But these are not infinity in the sense that they do not satisfy our intuitive desires for the properties that such an entity is supposed to have. Still, we would think that infinity ought to be so huge that adding one, or, for that matter, adding or subtracting any real number to it, would make no difference at all in its value. We might say that if we denote an entity I to be infinite, we should be able to prove that

$$I + r = I$$

for any real number r. Then we could also say that if R represents the numerical entity consisting of all the real number values,

$$I + R = I$$

It is clear that the entity R itself satisfies this requirement quite well. But we are looking for something else that might satisfy it.

Clearly, aleph-null can satisfy the requirement. If we let this transfinite cardinal be represented by the set of positive integers $\{1, 2, 3, 4, \ldots\}$, then adding one more element might result in the set of nonnegative integers $\{0, 1, 2, 3, \ldots\}$. These sets have the same cardinal number of elements, and therefore $\aleph_0 + 1 = \aleph_0$. From this it follows that $\aleph_0 + n = \aleph_0$ by mathematical induction, n being any member of the set of positive integers.

Subtraction of integers can be shown also to result in no change of the value of the transfinite cardinal $\aleph_0$.

Although I do not endeavor to formally prove it here, all of the transfinite cardinals satisfy the requirement that adding one results in no change in their value.

How about the mystical entity 1/0? What would happen to this if we add one to it? We might investigate this as follows:

$$1/0 + 1 = \frac{1}{0} + \frac{1}{1}$$
$$= \frac{1\times1 + 0\times1}{0\times1} \quad \text{(cross multiplication)}$$
$$= (1 + 0)/0 = 1/0$$

Thus, it appears that 1/0 satisfies the requirement of $I + 1 = I$ that we have intuitively supposed infinity should possess. It is not clear, though, that we can get away with cross multiplying if zero is in a denominator.

We can also demonstrate that the entity 0/0 satisfies the requirement. This can be investigated as follows:

$$0/0 + 1 = \frac{0}{0} + \frac{1}{1}$$
$$= \frac{0\times1 + 0\times1}{0\times1} \quad \text{(cross multiplication)}$$
$$= (0 + 0)/0 = 0/0$$

Again, there are zeroes in some of the denominators, and we have not established that it is "legal" to cross multiply under these circumstances.

The idea that 1/0 or 0/0 satisfy the requirement that adding one makes no difference in their value, is not anywhere near a conclusive proof that either entity is actually infinity. What we *can* say, according to entity theory, is that these expressions are numerical entities. They do exist, simply because it is possible to write the expressions down. They do not obey exactly the same laws of arithmetic as ordinary real numbers, but this is because they have multiple numerical values.

Georg Cantor showed that there are different degrees of infinity, by regarding the infinite as the number of elements in an infinite set. He did not attempt to postulate that division by zero might result in infinite numbers. When we investigate the properties of entities such as 1/0 and 0/0, we do not establish which, if any, of Cantor's "alephs" we are looking at. In fact, no distinction is clear from the above quasi-arithmetical manipulations.

You might object to the idea that there could be different degrees of the infinite. How can anything be greater than infinity? Yet it does follow that some infinities are greater than others; by talking simply about "infinity," we are not being specific as to which definition we are using. It matters how we approach the idea. Is infinity the number of natural, or counting, numbers? The number of rationals? Irrationals? Reals? It makes a difference.

The true general meaning of "infinity" is too nebulous to take these fine distinctions into account. It may be that 1/0 is, in some sense, equal to a numerical entity that has two or more "alephs" as values, but defining 1/0 goes no further than a very shallow answer to our curiosity about infinite sets.

Ordinal Numbers

All things in mathematics may be reduced in form to sets. In fact, the whole discipline of mathematics might be thought of as a specialty in the more general field of set theory. This is why set theory is taught at the beginning of almost every elementary-school grade, and it is used as the introductory material in many higher-level courses also. Set theory forms the foundation for modern mathematics. All numbers can be regarded as sets. The counting numbers are sometimes seen defined as sets such that each natural number n is contained in (is a member of) a set equivalent to $n + 1$. In fact, we may say that $m < n$, where m and n are natural numbers, if and only if m is contained in n. And if $m = n$, the sets are identical. This property of natural numbers is known as *well-ordering*.

How are numbers defined as sets? We begin by defining zero as the empty set, or the set without any elements in it defined as $\varnothing$. Then one is the set containing the empty set, written either $\{\varnothing\}$ or $\{\{\}\}$. Two is the set that contains zero and one, written $\{\varnothing, \{\varnothing\}\}$. Three is the set containing zero, one, and two. In general, the number n is the set containing the elements zero through $n - 1$ as previously defined. The natural numbers are thus built upon each other. The first m elements of a set representing a number n, where $m < n$, is simply the number m by definition. We call $I(m)$ the initial segment of the set representing a number n greater than m. So for example, if $n = \{0, 1, 2, 3, 4, 5, 6\} = 7$, and $I(4)$ is the initial segment of the first four elements of n, namely $\{0, 1, 2, 3\}$, then $I(4) = 4$. It is so simple as to be trivial. I have explained it this thoroughly only so that you might see how utterly simple it actually is, and that there is nothing whatsoever that is any more complicated about it. (It's as if to repeatedly emphasize, "Can you believe it?")

We define a set w as an *ordinal number* if w can be well-ordered so that for every element v of w, the initial segment $I(v)$ is equal to v. It is immediately clear that all of the positive integers are ordinal numbers. The set of all natural numbers, or positive integers (the latter does not include zero), in their natural order, is an ordinal number and is sometimes written using the Greek lower-case letter omega (ω). This letter is probably chosen because it is the last letter of the Greek alphabet and this number may be regarded as the last word in counting numbers—although it is far from the last word for all numbers generally. This ordinal will be written using the English lower-case letter z from now on herein, because it is easier to typeset and is less intimidating to look at. I'm not trying to impress anybody by using unnecessarily highfalutin' symbols.

The ordinal z is not finite, and we therefore refer to it as infinite or transfinite. The term *transfinite ordinal* arises from the fact that there are various levels of infiniteness, not just one, and that these levels are all less than the unreachable *absolute infinite*.

Theorems About Ordinal Numbers

The theorems here are simply stated and explained where necessary. The proofs may be found in a text about set theory, but total rigor and thoroughness is not the object here.

THEOREM 1. Every element of an ordinal number is an ordinal number; and every ordinal number is an element of some larger ordinal number. Recall that all numbers may be defined as sets. Then the numbers are built up as sets, each natural number n being an element of the next larger natural number $n + 1$. It follows that even the ordinal z, the infinity of the natural numbers, is an element of some ordinal larger than itself.

THEOREM 2. No ordinal is an element of itself. If an ordinal w is an element of some other ordinal v, then $w < v$. This theorem guarantees, with THEOREM 1, that there exist larger ordinals than z. In fact, there exist infinitely many ordinals larger than z.

THEOREM 3. For any two ordinal numbers v and w, it must be true that exactly one of the following holds: $v = w$, or v is equal to some initial segment of w, or w is equal to some initial segment of v. This is sometimes called the *trichotomy theorem*. It is equivalent, in number theory, to the trichotomy that for two numbers x and y, $x = y$ or $x < y$ or $x > y$.

THEOREM 4. An ordinal number is finite if it is a natural number, and the converse is also true: a number is a natural number if it is a finite ordinal. This means that the set of ordinals, as we build them up from the initial set $0 = \varnothing$, follows the sequence of what we think of as natural numbers 0, 1, 2, 3,.... There exist an infinite number of *finite* natural numbers, and each one of them is a unique finite ordinal.

Although all of this might sound formal and rigorous, keep in mind that it is so simple, if taken step by step, as to appear trivial. You might still recall those early-season (fall) school days in elementary and junior-high school where the teacher would talk about sets, and you, looking out at the still summer-green landscape, would try to make like you were paying attention. You might have wondered, then, why they would be teaching you things that were so obvious! An understanding of transfinite numbers is just one step beyond this elementary set theory.

THEOREM 5. Each ordinal w is equal to the set of all ordinals less than itself. Now this is not quite so trivial. But its proof is simple enough. If an ordinal x is less than w, then x is contained in w. If y is an element of w, then y is an ordinal and y is less than w. We may define, here, what it means for an ordinal x to be less than w: it simply means that x is an initial segment of w. Every initial segment of an ordinal w is an element of the set w.

THEOREM 6. Given an ordinal number w, its immediate successor is the ordinal that is the union of w and the set containing w: $S(w) = w \cup \{w\}$. Here, we write $S(w)$ to mean the successor of w. Examples are $5 = S(4)$ and $204 = S(203)$. The union of two sets is denoted by the symbol $\cup$. Every ordinal has a successor. This is as true of the first transfinite ordinal as it is of any other ordinal. Therefore $S(z) = z \cup \{z\}$, and this ordinal, $S(z)$, is larger than z. In fact, $z \cup \{z\}$ gives the exact definition of the first ordinal that is larger than aleph-null, the number of natural numbers.

Here we introduce the definition of *limit number*. A limit number is an ordinal that has no immediate predecessor. If the successor is denoted by $S(w)$ for some ordinal w, then w is the immediate predecessor of $S(w)$. It is clear that the only natural number that is a limit number is zero. This number has no immediate predecessor. You might want to think that -1 is a predecessor, in fact the immediate predecessor, of zero, but not according to the set-theory definition of natural numbers. The empty set has no elements, and $0 = \varnothing$. We cannot single out any elements in $\varnothing$, since there are none; so zero has no immediate predecessor. Every other natural number does. But, interestingly, the first transfinite ordinal, z, does not. We might try to name one. I'll leave that exercise up to you. But don't try it for too long. You can't do it.

The fact that z has no predecessor, or at least no immediate predecessor, has resulted in its definition as *not reachable from below*. There exists no ordinal number z^* such that $z^* < z$ and $z = S(z^*)$. In this sense, z is a sort of "aloof" ordinal. It sets itself apart from all of the natural numbers that comprise it.

Interestingly, the successor to z, $S(z) = z \cup \{z\}$, is not a limit number. This is simply because its predecessor is z, and this is an immediate predecessor.

THEOREM 7. There exists no set of all ordinal numbers. This might seem rather strange. How can the ordinals even exist if we cannot imagine a set that contains them all? There exists a set of all natural numbers, even though we cannot count through all of them nor actually envision the entire set at once. The proof of this theorem rests on the principle of *reductio ad absurdum*, which is assuming that there does exist a set of all ordinals and then showing that a contradiction results from this proposition. If there is a set of all ordinals called Z, then $Z \cup \{Z\}$ or $S(Z)$ is an ordinal, and it is greater than every ordinal belonging to Z. But since $S(Z)$ is an ordinal, it must belong to Z, the set of all ordinals. The ordinal cannot be greater than itself.

In a sense, we might think of the set of all ordinals—a set that does not actually exist except in wishful imagination—as a form of the absolute infinite. It can never be reached. It is impossible mathematically to collect all of the ordinals and put them in a "mind basket" and then say, "Here are all the ordinals." There would have to be at least one more that we overlooked.

Adding Ordinals Together

Numerous general addition facts have been proved for ordinal numbers. We will not go through the proofs here but will state some of the known addition facts. Proofs can be found in an advanced text on set theory.

THEOREM 8. For any ordinal number v, $0 + v = v + 0 = v$.

THEOREM 9. For any ordinal number v, $v + 1 = S(v)$.

THEOREM 10. For any finite ordinal v, $v + z = z$, where z is the first infinite ordinal. However, if u and v are two different finite ordinals so that $u \neq v$, then $z + u \neq z + v$. This is a rather bizarre result; suppose that $u = v + w$, where w is another finite ordinal. (The ordinals might be, for example, $u = 246$, $v = 147$, and $w = 99$.) Then we have $z + u = z + v + w$, but this is not equal to $z + v$. But it would seem that $w + z = z$ from the theorem, and thus $u + z = v + z$. The confusion arises from the fact that, in general, ordinal addition is not commutative.

THEOREM 11. In general, ordinal addition is not commutative. That is, it is *not* universally true that $u + v = v + u$ for all ordinals u and v. We might think of it as follows for the infinite case where one ordinal is z. If we add on a natural number of elements at the beginning of a denumerable set, say by changing the set $\{5, 6, 7, \ldots\}$ to $\{0, 1, 2, 3, 4, 5, 6, 7, \ldots\}$, we still have the same number of elements. But we cannot as readily envision how we would go about adding elements on the end of the set, the right-hand extreme to which we never get. Mathematicians have defined ordinal addition in such a manner so as not to render it commutative like ordinary addition. For example, we cannot say that $5 + z$, the result of adding the elements 0, 1, 2, 3, and 4 to the set $\{5, 6, 7, \ldots\}$, is the same as $z + 5$, where we would add five elements (not already in the set) to the right-hand extreme end.

THEOREM 12. Ordinal addition is associative. That is, if u, v, and w are any ordinals, then $(u + v) + w = u + (v + w)$. We may say, for example, that $u + z + v$, where u and v are finite ordinals, is equal to both $(u + z) + v$ and $u + (z + v)$. We know that $u + z = z$; therefore $z + v = u + z + v$ for any finite ordinals u and v.

THEOREM 13. For any two ordinals u and v, $u + S(v) = S(u + v)$.

THEOREM 14. For any two ordinals u and v, if $v > 0$, then $u + v > v$. This appears trivial, and indeed for finite ordinals it is. But for the infinite ordinal z, this theorem tells us that $z + n > z$ for any natural number n other than zero. This is not what we would expect for an infinite quantity. It is true that $n + z = z$, as we would expect, but since addition is not commutative, the fact that $n + z = z$ does not imply that $z + n = z$.

THEOREM 15. For any ordinals u and v, if $v > u$, then there exists a unique ordinal w, where $w \neq 0$ and $v = u + w$. This is again trivial for the case where u and v happen to be natural numbers. But either or both of the two could be infinite ordinals.

THEOREM 16. For any ordinals u, v, and w, $u + v = u + w$ if and only if $v = w$. One of the consequences of this is that $z \neq z + 1 \neq z + 2 \neq z + 3$ and so on. We in fact can say that $z < z + 1 < z + 2 < z + 3 \ldots$.

THEOREM 17. For any three ordinal numbers u, v, and w, $u + v < u + w$ if and only if $v < w$. Again, this would appear obvious from natural-number theory. But it is not necessarily true that $v + u < w + u$ if and only if $v < w$. Witness the case where $u = z$, and let $v = 3$ and $w = 5$. We have the fact that $3 + z = 5 + z$. However, $z + 3 < z + 5$.

Again, it is the noncommutative nature of ordinal addition that is the reason for this.

Sums of ordinals need not be finite. It is possible to define the sums of various infinite series of ordinals, even if some or all of the ordinals are themselves infinite. There is one theorem that especially deals with this.

THEOREM 18. A series of ordinal numbers is equal to the limit of the sequence of its partial sums. Here, some definitions are in order. A series is itself a sum of a sequence of numbers. We might have the sequence 1, 2, 3,... and its corresponding series is 1 + 2 + 3 +.... Clearly this is an infinite series. In this case, the sum is infinite. We could have the sequence 1/2, 1/4, 1/8,..., $1/(2^n)$,.... This suggests the series 1/2 + 1/4 + 1/8 +...+ $1/(2^n)$ +..., which is finite and is equal to 1. The partial sums are those sums from the first member of the series up to the *n*th member. For example, we may have the partial sum of the first four members of the above series as 1/2 + 1/4 + 1/8 + 1/16 = 15/16. The limit of the partial sums is the number 1, and this, according to the theorem, is the sum of the whole infinite series.

The above theorem allows us to evaluate many infinite series that have sums equal to the infinite ordinal *z*. For example, if the partial sum of a series is always a natural number, and this value always gets larger as we take an additional term, then the sum is *z*. Examples of such series are as follows.

```
1  +  1  +  1  +  ...        (series)
|     |     |
1     2     3     ...        (partial sums)
1  +  2  +  3  +  ...        (series)
|     |     |
1     3     6     ...        (partial sums)
0  +  2  +  4  +  ...        (series)
|     |     |
0     2     6     ...        (partial sums)
```

The first series contains all numbers 1. The second series contains numbers that increase by one each time; the last series consists of the even natural numbers, beginning with zero. In each case, the partial sum is always a natural number, as long as the number of terms in the series is finite. But the limit is not within the natural numbers; the series diverge in each of these cases. The sum is therefore *z*, the infinity of the natural numbers, because it is a countable infinity.

Multiplying Ordinals

Ordinal multiplication is defined, as is ordinal addition, according to the concept of the ordered product of well-ordered sets. Ordinal multiplication is noncommutative; that is, in general it is not true that $uv = vu$, where u and v are ordinals. For natural numbers, this property does hold, but for cases where one or both ordinals are infinite, the property breaks down. In general, multiplying two ordinals u and v is defined as u times v, or u added to itself v times. This means that $2z = z$, but $z2 = z + z \neq z$.

THEOREM 19. Ordinal multiplication is associative. That is, for any three ordinals u, v, and w, $u(vw) = (uv)w$. We may therefore write uvw and not be ambiguous. An interesting fact arises from this theorem. If, for example, $z3z = z(3z) = (z3)z$, but $z(3z) = zz$, which we may call z^2, then $(z3)z = (z + z + z)z$. Thus, $z^2 = (z + z + z)z$.

THEOREM 20. Ordinal multiplication is left distributive with respect to ordinal addition. That is, for any three ordinals u, v, and w, it is true that $u(v + w) = uv + uw$.

THEOREM 21. For every ordinal u, $0u = u0 = 0$.

THEOREM 22. For every ordinal u, $1u = u1 = u$.

THEOREM 23. For any two ordinals u and v, $uv = 0$ if and only if $u = 0$ or $v = 0$.

THEOREM 24. For any three ordinals u, v, and w, if $u > 0$ and $v > w$, then $uv > uw$.

THEOREM 25. For every four ordinals u, v, w, and x, if $u \leq v$ and $w \leq x$, then $uw \leq vx$.

THEOREM 26. For any two ordinals u and v, if $u > 0$ and $v > 0$, then $uv \geq u$, and also $uv \geq v$.

THEOREM 27. For any two ordinals u and v, if $u > 0$ and $v > 1$, then $uv > u$.

THEOREM 28. For any three ordinals u, v, and w, if $uv < wv$, then $u < w$.

THEOREM 29. For any three ordinals u, v, and w, if $u > 0$ and $uv = uw$, then $v = w$.

There are many other theorems along this line; stating many more would become rather tedious. In many cases, the theorems seem trivial, as the facts are well known for natural numbers. It is important to remember that the order of multiplication is important when considering ordinals in general, because if either or both of the ordinals is infinite, the operation is noncommutative.

Exponents with Ordinals

With ordinal numbers, as with ordinary numbers, exponentiation is defined as repetition of multiplication. For example, the expression u^v is the product of the ordinal u with itself, v times. That is, $u^v = uuu...u$, where "u" is written v times. This product may be finite; this will always be the case when u and v are finite. But v need not be finite. It is entirely possible to have infinite products just as it is possible to have infinite sums.

We might begin by talking briefly about infinite products. In order to define an infinite product, we define the partial products of a sequence of ordinals. When we can define the partial product of a sequence of positive ordinals, then we can define the infinite product as the limit of the sequence of partial products.

For natural numbers, we might have

$$\begin{array}{ccccccccc} 0 & \times & 0 & \times & 0 & \times & \ldots & & \text{(sequence)} \\ | & & | & & | & & & & \\ 0 & & 0 & & 0 & & \ldots & & \text{(partial products)} \end{array}$$

$$\begin{array}{ccccccccc} 0 & \times & 1 & \times & 2 & \times & \ldots & & \text{(sequence)} \\ | & & | & & | & & & & \\ 0 & & 0 & & 0 & & \ldots & & \text{(partial products)} \end{array}$$

$$\begin{array}{ccccccccc} 1 & \times & 1 & \times & 1 & \times & \ldots & & \text{(sequence)} \\ | & & | & & | & & & & \\ 1 & & 1 & & 1 & & \ldots & & \text{(partial products)} \end{array}$$

$$\begin{array}{ccccccccc} 1 & \times & 2 & \times & 3 & \times & \ldots & & \text{(sequence)} \\ | & & | & & | & & & & \\ 1 & & 2 & & 6 & & \ldots & & \text{(partial products)} \end{array}$$

The limit of the partial products in the first three cases above is a finite ordinal: zero in the first two cases and one in the third case. In the fourth case, the partial products are ever-larger natural numbers, so the limit is z, the ordinal that represents the infinity of the natural numbers.

The above sequences, and consequently the products, contain z terms each. This is simply because the terms are infinite in number, and we can count them (although we cannot count all the way to the end of the sequences). In an exponential sense, then, the above expressions represent 0^z for the first expression and 1^z for the third expression. (The second and fourth expressions are not reducible in this way because the product is not one of all the same term.) We can therefore say that $0^z = 0$ and $1^z = 1$.

Let us now allow, for a moment, the base of an exponential expression to be something other than an ordinal—let it be any real number. What happens if we raise some real number r to the power z? We will then have the expression

$$\begin{array}{ccccccccc} r & \times & r & \times & r & \times & \ldots & & \text{(sequence)} \\ | & & | & & | & & & & \\ r & & r^2 & & r^3 & & \ldots & & \text{(partial products)} \end{array}$$

Consider first the case where $0 < r < 1$. Then the partial products go toward zero, and zero is in fact the limit product. The closer r is to zero to begin with, the faster it will approach zero—but even 0.99999999 will eventually converge toward zero. A rigorous proof of this is not given here, but it should be fairly easy to grasp. Taking a fraction of any number will yield a smaller number, that is, a number closer to zero, if that fraction is between (not including) 0 and 1. Thus, if $0 < r < 1$, $r^z = 0$ (Fig. 4-2).

Now consider the case $r > 1$. This will invariably result in a series of increasing real numbers. In the special cases where r is a natural number, we will obtain an increasing

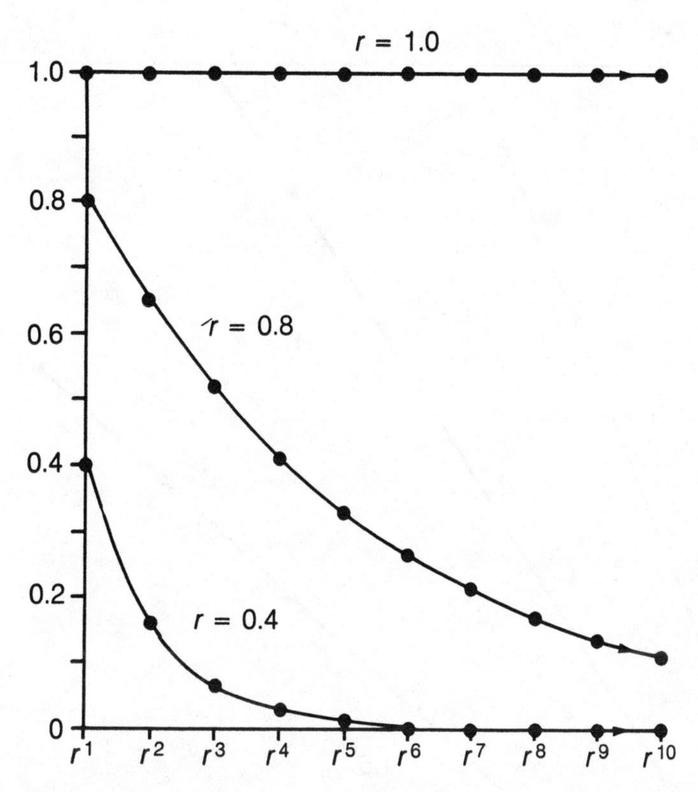

Fig. 4-2. Exponentiation of real numbers $r = 1.0$, $r = 0.8$, and $r = 0.4$. Note the convergence toward zero for r less than 1.

series of natural numbers. In any case, we can count the terms even though we can't get to the end of the sequence; so, if $r > 1$, then $r^z = z$ (Fig. 4-3).

It is more difficult to evaluate the real numbers r such that $r < -1$. In the case $-1 < r < 0$, the product converges toward zero but becomes alternately positive and negative; hence in this case $r^z = 0$. But when $r < -1$, the product grows farther and farther from zero, alternating positively and negatively. We might say then that if $r < -1$, $r^z = (z, -z)$ the entity of both z and $-z$. The situation for $-1 < r < 0$ is shown at Fig. 4-4, and the situation for $r < -1$ is shown at Fig. 4-5.

Obviously, the behavior of ordinals as exponents is a strange business, unlike anything in finite mathematics. There are some theorems we can state concerning exponentiation of, and with, ordinals.

THEOREM 30. For any ordinal u, $u^1 = u$ and $1^u = 1$.

THEOREM 31. For any two ordinals u and v, if $u > 0$, then $u^v > 0$.

THEOREM 32. For any three ordinals u, v, and w, if $v < u$ and $w > 1$, then $w^v < w^u$.

THEOREM 33. For any three ordinals u, v, and w, if $w > 0$ and $v \leq u$, then $w^v \leq w^u$.

THEOREM 34. For any three ordinals u, v, and w, if $w > 1$ and $w^u = w^v$, then $u = v$.

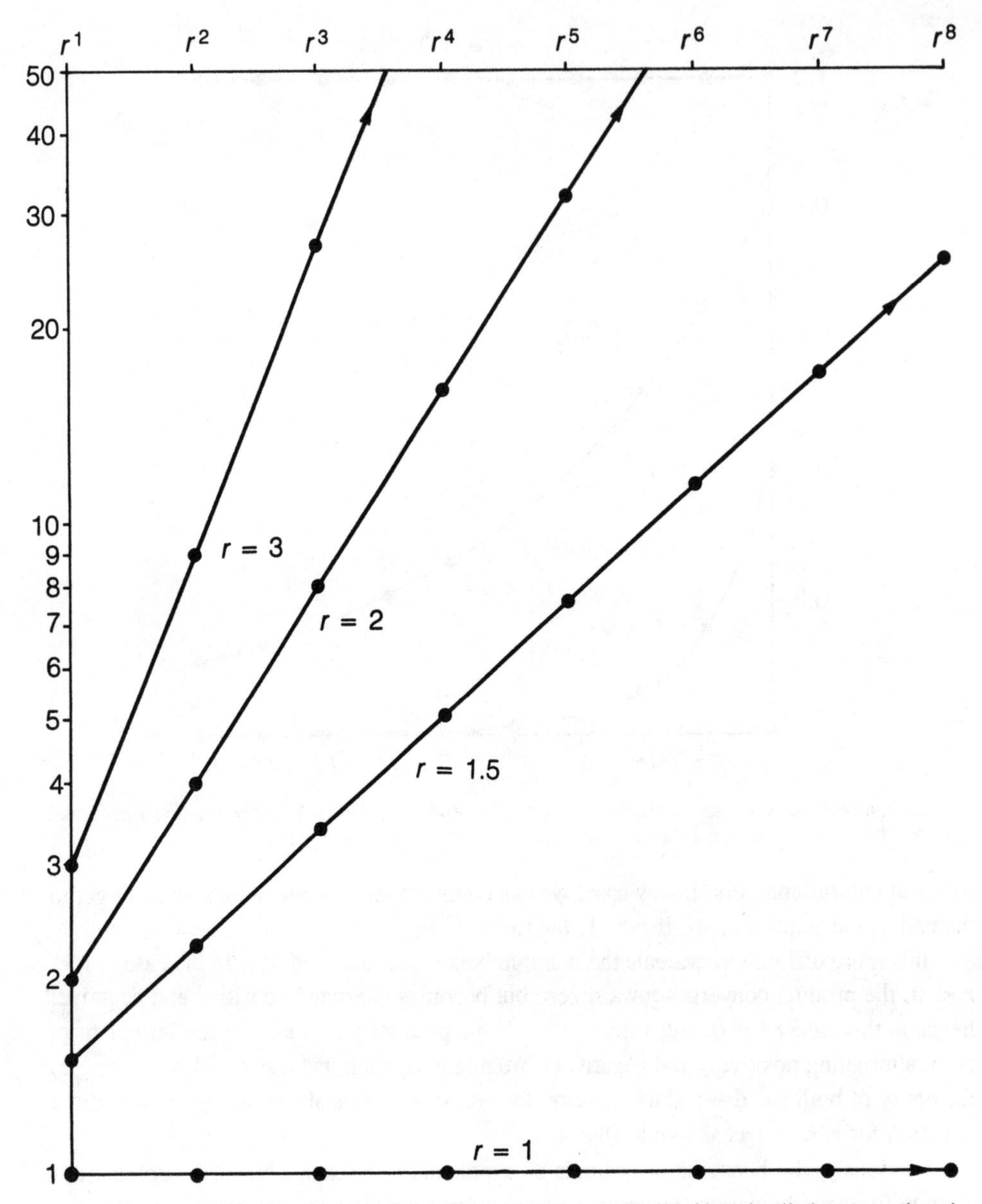

Fig. 4-3. Exponentiation of real numbers $r = 1$, $r = 1.5$, $r = 2$ and $r = 3$. Note that if r is greater than 1, the value increases without bound.

THEOREM 35. For any three ordinals u, v, and w, if $w > 1$ and $w^u < w^v$, then $u < v$.

THEOREM 36. For any three ordinals u, v, and w, $w^u w^v = w^{(u+v)}$.

THEOREM 37. For any three ordinals u, v, and w, $(w^u)^v = w^{(uv)}$.

THEOREM 38. For any two ordinals u and v, if $u > 1$, then $u^v \geq v$.

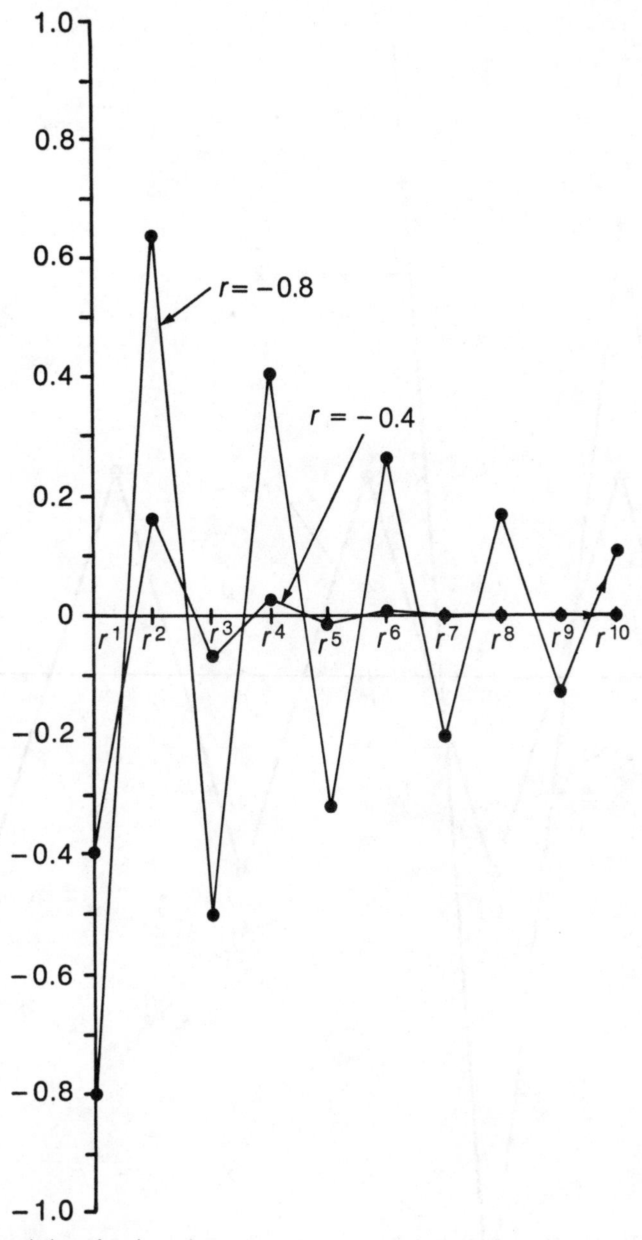

Fig. 4-4. Exponentiation of real numbers $r = -0.4$, $r = -0.8$. Note the convergence toward zero with alternating positive/negative values.

THEOREM 39. For any three ordinals u, v, and w, if $u \leq v$, then $u^w < v^w$.

THEOREM 40. If n is some finite positive ordinal, then for any two ordinals u and v, if $uv = vu$, then $u^n v^n = v^n u^n$. The converse also holds; if $u^n v^n = v^n u^n$, then $uv = vu$.

THEOREM 41. If u, v, and w are ordinals and $u > 1, uv = vu$ and $uw = wu$, then $vw = wv$.

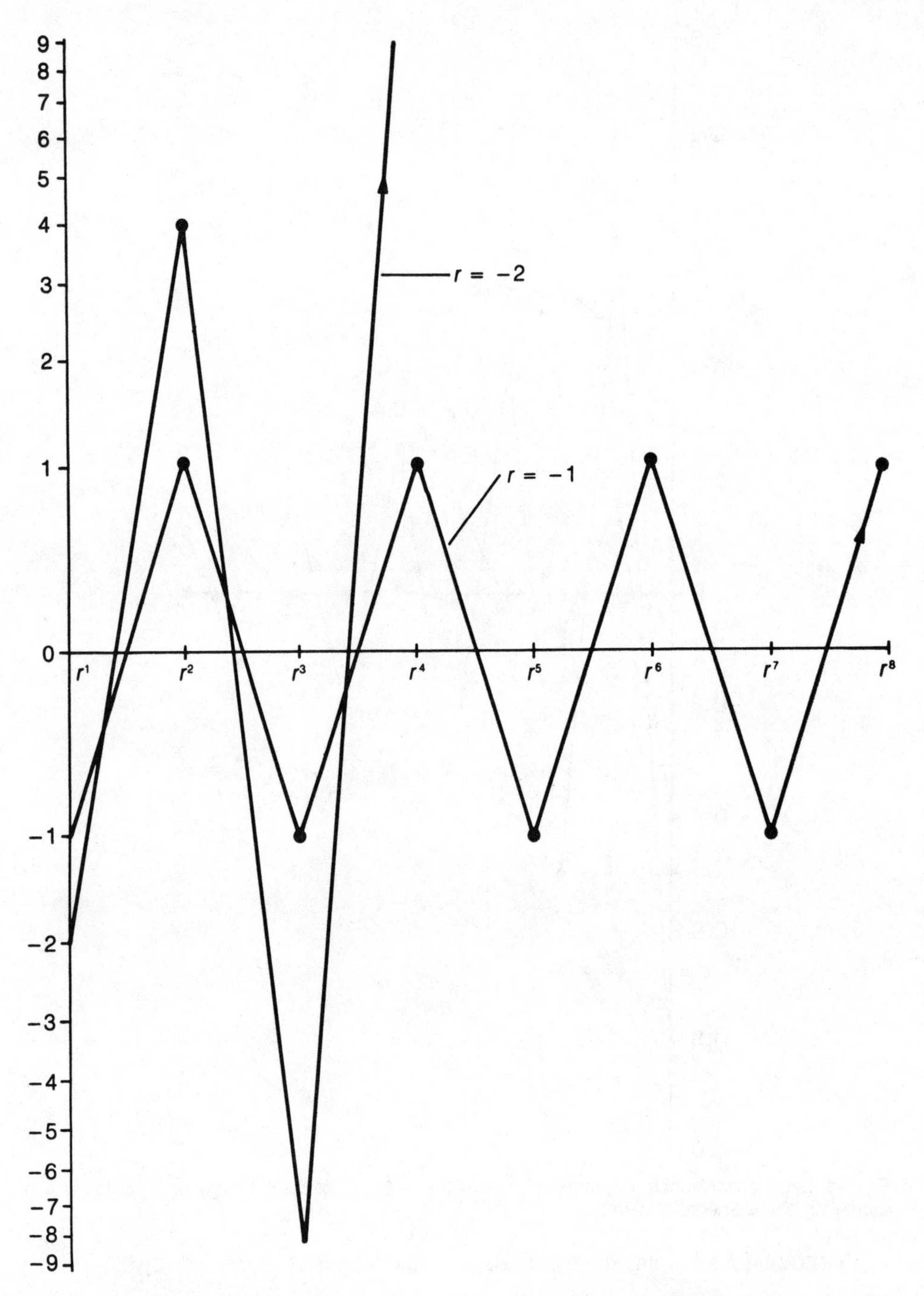

Fig. 4-5. Exponentiation of real numbers $r = -1$, $r = -2$. Note the values always alternating in sign, but diverging without bound if r is less than -1.

The Hierarchy of Infinite Ordinals

We have seen some of the facts concerning addition, multiplication, and exponentiation of ordinals. We have defined z as the ordinal number that represents the infinity of the counting numbers. (Remember that it is often called *omega* and written as the lower-case Greek symbol ω, but I have avoided this notation to make the whole business look a little bit less intimidating.)

Each time we perform some act on z that results in an ordinal greater than z, we have reached a larger infinity, some higher cliff on a mountain with infinitely many cliffs. There is in fact a hierarchy of infinities if we accept the ordinal definition. There is no end to the levels of this hierarchy.

An interesting thing about this hierarchy of infinities is that it is comprised of discrete entities. That is, for each specific ordinal in the hierarchy, there exists a unique ordinal that is the immediate successor of the ordinal in question. Therefore, $z + 1 > z$ and is the immediate successor of z because there is no ordinal smaller than $z + 1$ and yet larger than z. Then $z + 2 = z + (1 + 1) = (z + 1) + 1 > z + 1$, and we can begin building a hierarchy $z < z + 1 < z + 2 < z + 3 < \ldots < z + n < \ldots < z + z$. Once we have reached $z + z$ or $z2$, we can begin anew with $z2 < z2 + 1 < z2 + 2 < z2 + 3 < \ldots < z2 + n < \ldots z2 + z = z3$.

This is by no means even a good start to what follows! For we can keep going on until we ultimately arrive at zz or z^2. But all we need do is add 1 to the right-hand side of this to obtain an ordinal larger still, and the whole process begins all over, culminating in $z^2 + z$; infinitely many more steps lead to $z^2z = z^3$. We of course have $z^4 < z^5 < z^6 < \ldots < z^z$, with many infinities in between each of these entities. Yet it, too, is only the tip of the proverbial iceberg—no—just a molecule on the tip! A quark in that molecule. An elementary particle, a black-hole universe.

The parallel with fundamental particles (with repeating hierarchies) and chaos theory (where infinitely complicated patterns repeat infinitely many times on an infinitude of scales) ought to be apparent here. We may attempt to look at the hierarchy of infinite ordinals with a mental "macroscope," a device that allows us to see infinitely large things, infinitely large collections of infinitely large things, and so on. But we will always see the same unending pattern of larger and larger infinities with no end to any of it on any scale.

Cardinal Numbers

Another way of approaching the concept of infinity is to use the idea of a cardinal number. A cardinal number, like an ordinal number or any number, is reducible in form to a set. Cardinal numbers deal with the size of a set. A general definition for the term is that a cardinal number is the number of elements in a set. Thus, the empty set has cardinality zero; a set with one element has cardinality of one. The set of natural numbers has infinite cardinality, as does the set of rationals, the set of irrationals, the set of reals, and the set of complex numbers. The set of points on a line has infinite cardinality, as does the set of points (a,b) in two-space where a and b are natural numbers. The list goes on and on, infinitely.

Every cardinal number is also an ordinal number, and the sum and product of any two cardinals is again a cardinal. We specifically define a cardinal number as an *initial ordinal*. This means that it is the first number of some class of numbers.

At this point, we also define the term *equipollence*. Here, I am avoiding too much formality so that this does not turn into an axiomatic set-theory dissertation. Equipollence basically means *equivalent in effect*. When we speak of set size, sets are equipollent if and only if they have the same number of elements. We can state a fundamental theorem about cardinals using this term:

THEOREM 42. Every set is equipollent to a unique cardinal number. This means that for any given set, the set is equivalent in behavior (effect) to one, and only one, cardinal number. The set of all stars in our galaxy is perhaps equipollent to the cardinal number 118,603,229,557, for example.

The mathematician Georg Cantor is credited for having first formulated a comprehensive theory of cardinal numbers. It was probably his objective to explore the infinite using this tool, as I have employed my little theory of multiple-valued numbers to the same end. Cantor got more satisfaction from his endeavor than I have gotten from mine; I managed only to prove that $1/0 = 0$. This is practically meaningless and worthless. Despite Cantor's fame today, he was, in his own time, scorned by the establishment. It is said that this drove him mad.

THEOREM 43. The ordinal z is equal to the cardinal aleph-null ($\aleph_0$), representing the number of elements in the set of natural numbers.

Theorems About Cardinal Numbers

We now state, without proof, some theorems about cardinal numbers. We first define the less-than-or-equal-to ($\leq$) operation for cardinals. For any two cardinals a and b, $a \leq b$ if and only if $a = b$ or a is equal to some initial segment of b.

The meaning of the term *initial segment* should be intuitively clear but is illustrated in Fig. 4-6. In a denumerable set, that is, in a set in which we can count off the elements (even if it is not possible to count all the way through them in a finite time), an initial segment consists of the first n elements where n is a natural number.

THEOREM 44. For any two cardinal numbers a and b, one of the following must hold true: $a = b$, $a < b$ or $b < a$. This is sometimes called the *law of trichotomy*.

We define a cardinal as *transfinite* if and only if it has an infinite number of elements; that is, if it denotes a nonfinite set.

THEOREM 45. There exist infinitely many transfinite cardinals, and they may be denoted by the initial sequence $\aleph_0 < \aleph_1 < \aleph_2 < \ldots$.

THEOREM 46. If we call the cardinal of all real numbers aleph-r ($\aleph_r$) then $\aleph_0 < \aleph_r$.

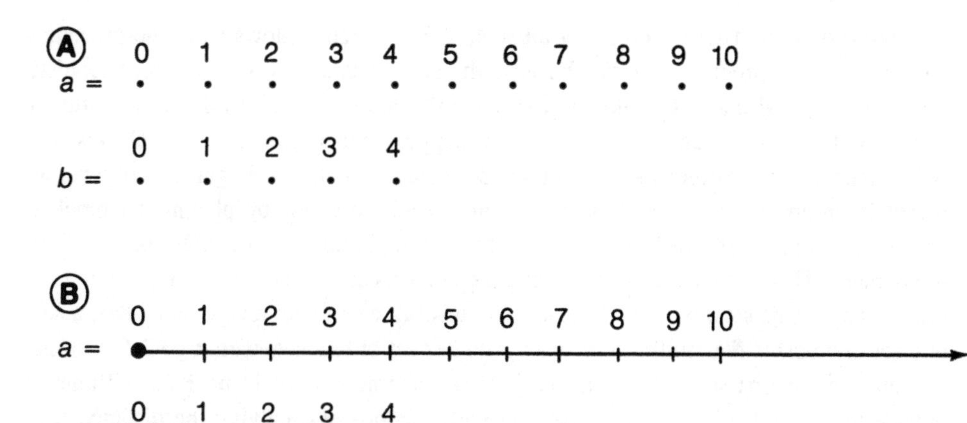

Fig. 4-6. Examples of initial segments. At (A), for discrete sets, *a* and *b* are both finite; at (B), *a* is the set of nonnegative reals, and *b* is the set of reals between 0 and 4. In each case, *b* is an initial segment of *a*.

The hypothesis that $\aleph_r = \aleph_1$ is called the *Continuum Hypothesis*. This hypothesis can be assumed true without contradiction. Georg Cantor was partial to the belief that the hypothesis is true, but he couldn't prove it. Thus, he believed the set of points on a line is of the same cardinality as the set of real numbers; that is, that the line is a true representation of the set of real numbers. If this is in fact the case, then we are indeed constructing a realistic model when we assign the set of real numbers to the points on a line. There aren't any "extra points" that have no real-number correspondent; also, there aren't any extra real numbers that have no place on the line. Unfortunately, the truth of this hypothesis has never been proven rigorously. It is possible also to assume that it is not true. If the Continuum Hypothesis is assumed not true, then either there are extra real numbers that have no place on the line (and this is extremely hard to believe), or there are more points on a line than can be accounted for by all of the real numbers in the real-number continuum. This is at least a plausible idea, although it is impossible to diagram and requires a certain frame of mind to even begin to envision.

THEOREM 47. There is no set of all cardinal numbers. This might come as a shock. At first it would seem that there ought to be as many cardinals as there are natural numbers; we might have $\aleph_n$ for $n = 0$, $n = 1$, $n = 2$, and so on. However, we may have "aleph-aleph-null," written

$$\aleph_{\aleph_0}$$

and this will be larger than any $\aleph_n$ with n a natural number. The situation becomes similar to that we saw with ordinals. We can construct ever larger cardinals, and we are limited only by our ingenuity and the time required to put down what we think of. As with ordinals, we encounter the concept of an Absolute Infinite, a thing that is larger than any cardinal. This implies that the Absolute Infinite is not a cardinal. It is perhaps what we think of when we think of "infinity."

THEOREM 48. There is no set of all sets. This theorem follows from the preceding theorem. Precise proof is not given because the background of theory is rather sophisticated and would also tend to make this discussion "watered down." Basically, an intuitive quasi-proof of this theorem can be obtained by supposing that there is a set of all sets; call is S. Then we can construct $\{S\}$, but this cannot be a member of S. Regardless of what we invent in an attempt to class all sets, we can create more sets by placing set brackets around it. If one set of brackets is not enough, we might put two, or three, or even infinitely many. This might be done by placing the brackets at ever-closer spacings from each other. Imagine the set S with the first set of brackets a centimeter away; then place a second set of brackets five millimeters away. Another set of brackets would go 2.5 millimeters outside the first set of brackets. A third set of brackets would be 1.25 millimeters outside the second set. We would keep on with this process, halving the distance each time, after the fashion of the IC (infinite coordinate) system axes. We might get an inscription like the illustration of Fig. 4-7. We could get around the problem of having only a finite amount of time to draw infinitely many brackets by making the actual brackets half as large with each succeeding repetition, thus finitizing the travel distance of the pen we use. Oh, but I hear someone saying, "How about the problem of fineness of the pen point?" We make it infinitely fine. "How can you do that," someone asks, "when atoms are only so small?" to which I reply, "How do you know there is a smallest particle at all?" Ultimately we can disregard all of these practical problems by reminding ourselves that we need not write anything to have a set; it exists already whether we write it or not.

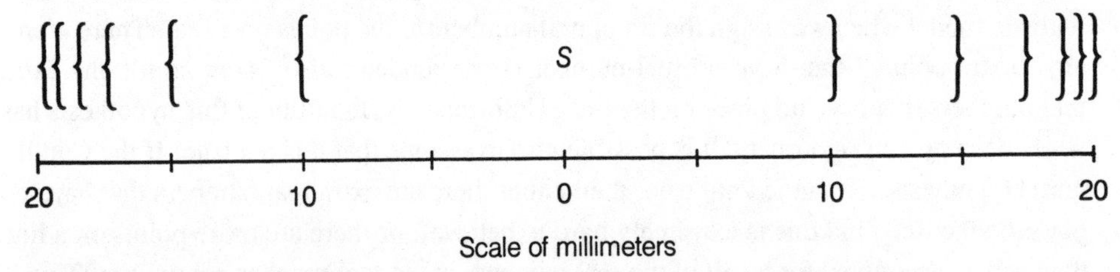

Fig. 4-7. A set with an infinite succession of brackets. Each succeeding pair of brackets is placed half the distance outside the previous pair.

THEOREM 49. There is no infinite decreasing set of cardinals. This means that if we begin with a certain cardinal and work our way downward, we cannot go on infinitely. We may think that we ought to be able to find a cardinal such that this is possible. How about aleph-null? We cannot even go one cardinal down from this; there is no next smaller one. No natural number will fill this requirement; there is always a larger natural number $n + 1$ for any large natural number n. This brings up an interesting characteristic of certain cardinals. Some are "approachable from below," while some are not.

Let us define a cardinal as being *approachable from below* if and only if there exists a smaller cardinal, such that there are no cardinals in between the two. Putting it specifically, a cardinal c is approachable from below if and only if there exists a unique cardinal c_*, such that $c_* < c$ and there is no cardinal d such that $c_* < d < c$.

THEOREM 50. All natural numbers, except zero, are approachable from below. This is simple enough to see; given a natural number n where $n > 0$, we may take $n - 1$, and we have $n - 1 < n$, and also the fact that there is no natural number m such that $n - 1 < m < n$. The reason that zero does not satisfy the requirement is that there is no natural number smaller than zero.

THEOREM 51. Aleph-null is not approachable from below. Suppose that it were. Then there would exist some cardinal c, such that $c < \aleph_0$ but also such that no cardinal d would satisfy $c < d < \aleph_0$. There are no such transfinite (infinite) cardinals smaller than $\aleph_0$. But there are certainly many cardinals that are finite and smaller than $\aleph_0$. They are, to be precise, the natural numbers. However, none of the natural numbers is such that it is the largest natural number. We can always find a natural number larger than any we care to name, simply by adding one. This shows that there can be no cardinal c that fills the above requirement. The contradiction proves the theorem by *reductio ad absurdum.*

THEOREM 52. There exists no greatest cardinal. This is similar to the assertion that there exists no greatest natural number. We can always show that $2^c > c$ for any cardinal number c. This is sufficient to prove that no cardinal can be the greatest cardinal.

Adding, Multiplying, and Exponentiating Cardinals

The rules for adding, multiplying, and exponentiating cardinals are different from the rules for ordinals. For transfinite (infinite) cardinal numbers, the rules appear to be entirely new. We know that $z + z \neq z$, but it turns out that $\aleph_0 + \aleph_0 = \aleph_0$, even though, curiously, $\aleph_0 = z$. The catch comes when we realize that, although we are speaking of the same magnitude of infinity when we speak of z and of $\aleph_0$, the operations of addition are different for the two types of entities. The ordinal z is what we would get if we could count all the way to the end of the sequence 0, 1, 2, 3, 4,...; the cardinal $\aleph_0$ is the number of elements in the set N of natural numbers $N = \{0, 1, 2, 3, 4,...\}$.

THEOREM 53. For any natural number n, $\aleph_0 + n = n + \aleph_0 = \aleph_0$.

THEOREM 54. $\aleph_0 + \aleph_0 = \aleph_0$.

THEOREM 55. For any two transfinite cardinals k_1 and k_2, if $k_1 < k_2$ then $k_1 + k_2 = k_2 + k_1 = k_2$. Thus, if we have two different "infinities" and one is larger than the other, when we add them, we end up with the larger of the two "infinities." Thus, for example, $\aleph_1 + \aleph_0 = \aleph_1$.

THEOREM 56. If k is a finite cardinal and k_1 is a transfinite cardinal, then $k + k_1 = k_1 + k = k_1$.

THEOREM 57. For every two ordinal numbers u and v, $\aleph_u + \aleph_v = \aleph_v$ if $u \leq v$.

THEOREM 58. If we let the power of the continuum be represented by $\aleph$, then $\aleph_0 + \aleph = \aleph_1 + \aleph = \aleph$. This means that the power of the continuum is at least as large as $\aleph_1$.

Multiplication of cardinal numbers is similar to the multiplication of ordinals in certain ways. However, it is best to imagine that the product of two cardinals is the number of elements in the two-space generated by taking the cardinals as axes in a coordinate system. This might be shown for the cardinal $\aleph_0$ by the drawing of Fig. 4-8. We end up with a grid of points that can be denumerated in various systematic ways. This is how we can tell that the following theorem is true:

THEOREM 59. $\aleph_0 \aleph_0 = \aleph_0$.

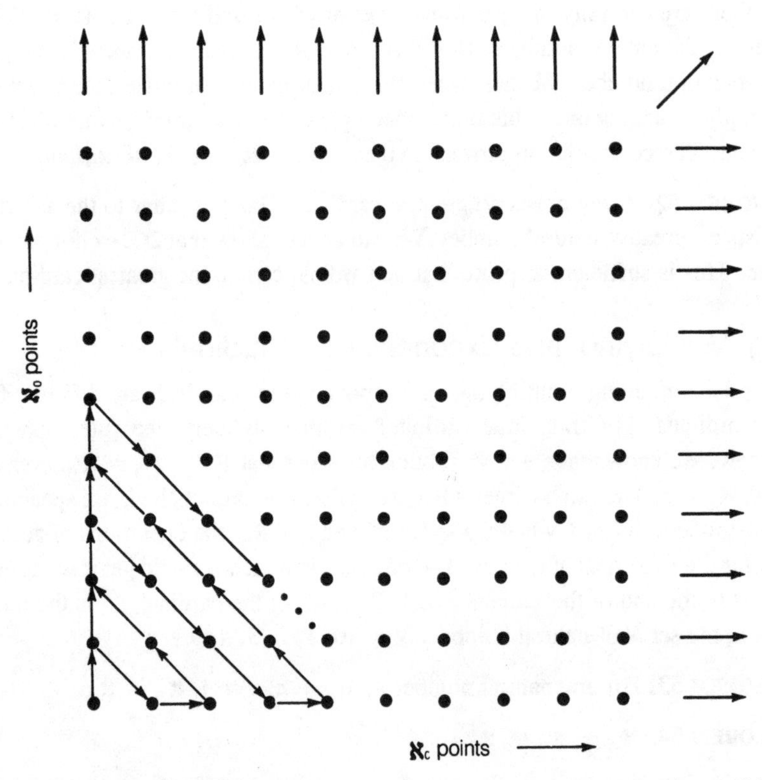

Fig. 4-8. Multiplication of two denumerably infinite sets, which is depicted by generating a grid of points, results in a set that is still denumerably infinite.

You might be helped in the envisioning of the product of cardinals if you consider the definition for finite cardinals. Say we have $a = 4$ and $b = 6$; the product is obtained by laying out a grid of points in a 4-by-6 array, thus obtaining 24 points. We would verify this number of points by actually counting them. Thus, we verify the number of points in the product of two cardinals by counting (or finding out that we cannot count) them.

THEOREM 60. If c is a finite cardinal other than zero, then $c\aleph_0 = \aleph_0 c = \aleph_0$.

THEOREM 61. For any cardinal numbers a, b, c, and d, if $a < b$ and $c < d$, then $ac < bd$.

THEOREM 62. For any nonzero cardinal a and any infinite cardinal b, if $a \leq b$, then $ab = b$.

The list can go on and on. The gist of cardinal addition and multiplication is that for any two cardinals a and b, if one or both are infinite, then the sum and product are equal to the larger of the two cardinals. We cannot obtain larger and larger cardinals, therefore, just by adding or multiplying many of them together. For $\aleph_0 \aleph_0 = \aleph_0$; we can show by induction therefore that $\aleph_0{}^n = \aleph_0$ for any natural number n. We must resort to exponentiation, in which the exponent itself is a transfinite cardinal, if we are to have any hope of generating larger and larger cardinals.

THEOREM 63. For any three cardinal numbers a, b, and c,

$$a^{(b+c)} = a^b a^c$$
$$(ab)^c = a^c b^c$$
$$(a^b)^c = a^{(bc)}$$

THEOREM 64. For any four cardinals a, b, c, and d, if $a \leq b$ and $c \leq d$ then $a^c \leq b^d$.

THEOREM 65. If a is a cardinal and $a > 1$ and b is an infinite cardinal such that $a \leq b$, then $a^b = 2^b$.

THEOREM 66. The number of infinite subsets of a denumerable set is the power of the continuum. This means that if we consider all of the infinite subsets of the set N of natural numbers, we will have more than $\aleph_0$ of these subsets. In fact, the number of subsets will be the same as the number of points on a line. The Continuum Hypothesis would have us believe that this is $\aleph_1$. But, as we have seen, this can only be conjectured. We might begin to try to denumerate the number of infinite subsets of N by multiplying each member of N by some natural number m. Then we would have sets that we might call N_1, N_2, N_3, and so on, as follows.

$$N_1 = \{0, 1, 2, 3, \ldots\} = N$$
$$N_2 = \{0, 2, 4, 6, \ldots\}$$
$$N_3 = \{0, 3, 6, 9, \ldots\}$$
$$N_m = \{0, m, 2m, 3m, \ldots\}$$

But this set of N_m, while an infinite (denumerable) set, is not the whole set of infinite subsets of N. We can strike out the zero, for example, in each of these, and get still more. The theorem states that we cannot ever hope to list all of the infinite subsets of N. We cannot even devise a scheme for denumerating them.

THEOREM 67. For every four cardinals a, b, c, and d, if $a^c < b^c$ then $a < b$. Also, if $a^c < a^d$ then $c < d$.

THEOREM 68. $2^{\aleph_0}$ is equal to the power of the continuum, and this is larger than $\aleph_0$.

We remember again that in this discussion, the symbol $\aleph$ refers to the power of the continuum, which according to the Continuum Hypothesis ought to be $\aleph_1$.

THEOREM 69. The following statements hold:

$1^n + 2^n + 3^n + \ldots = \ {}_0$ for n, a natural number

$1^n 2^n 3^n \ldots =$ for $n > 0$

$n^1 + n^2 + n^3 + \ldots = \ {}_0$ for $n > 0$

$n^1 n^2 n^3 \ldots =$ for $n > 1$

Interestingly, for the ordinal z in ordinal arithmetic, $2^z = z$. But we have seen that for aleph-null in cardinal arithmetic, this does not hold true.

The Absolute Infinite

We have seen that there are infinitely many transfinite cardinals (Theorem 45) and ordinals. We can keep making them bigger and bigger. This reopens the question, "Is there an infinity that cannot be reached in any sense?" No matter how many millions of centuries we spend writing down notations with aleph symbols or ordinal symbols, we cannot reach such a limit. We cannot reach it even in $\ {}_0$ years! Clearly there must indeed be such a thing as an Absolute Infinite, something that cannot be reached in any sense, not even by trying for infinitely long to invent new and better ways of infinitely accelerating the inscription process. No matter how hard we stretch the imagination, there is no way we can hope to achieve all that is.

And yet, merely by breathing the words "Absolute Infinite," have we not conceived of exactly that which we have said we cannot conceive?

We could try to apply the same argument, in an attempt to escape this issue, as we apply to the Berry Paradox that says we cannot hope to write down or utter the smallest natural number larger than anything we can write down or utter. Yet, such a number must exist, according to the Least Upper Bound theorem, and by saying just that, we have identified it, even if not precisely in a numerical fashion. But that befuddlement is just a weak attempt at obtaining a true paradox, compared with the paradox of the Absolute Infinite. The very definition of such a thing implies that it cannot be conceived; yet we need only acknowledge that the thought has occurred to us in order to conceive it and defy its own definition.

We might legitimately ask whether or not all of the alephs or ordinals actually exist. They are numbers—not finite numbers, but a different breed. We have no problem with the existence of finite numbers as inventions for our convenience. The number two, for example, cannot be touched or blown up or used for anything at all except to express a certain notion; the same is true for aleph-two. So of course the alephs and ordinals all exist within the structure of mathematics that we accept for our use.

Chapter 5

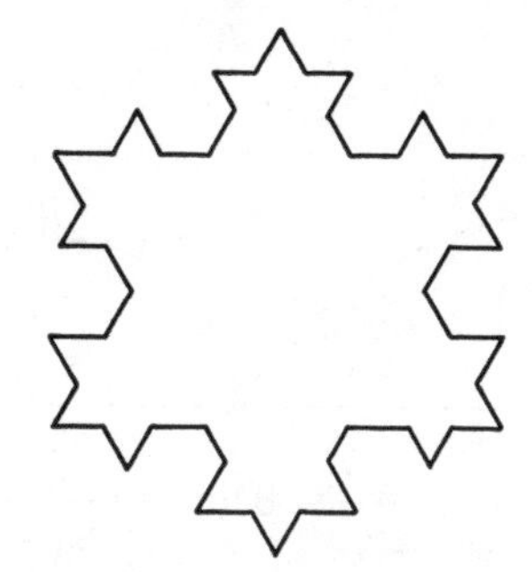

Bringing Infinity into View

MANY OF THE PRINCIPLES USEFUL IN UNDERSTANDING "INFINITY" ARE TAUGHT IN elementary school. The problem is that we forget them, even though they are taught over and over, usually at the beginning of each school year. Elementary point-set geometry and set theory are fundamental to all of mathematics. The pursuit of infinity is no exception.

The Frog and the Wall

Suppose we have a frog and a wall a certain distance away from the frog—8 meters, for example. Now imagine the frog jumping exactly half the way to the wall. This is a 4-meter jump, quite a leap for most frogs, but this frog is a perfect frog, constructed in the imagination. It could leap 400 meters without going a millimeter more or less. So it jumps precisely 4 meters the first time.

Then it jumps halfway to the wall again, or exactly 2 meters, and again—1 meter. It keeps on repeating this halfway-to-the-wall jump: first 500 millimeters, then 250, then 125, and so on, without end. We could number each jump integrally so that jump number 1 is 4 meters, number 2 is 2 meters, and so forth (see Fig. 5-1A).

Will the frog reach the wall? Not in a finite number of jumps, it won't. It can keep jumping forever, say at the rate of one jump per second, and it will never reach the wall. It will still be jumping after a day, a month, a year, a century. (It is an immortal frog so it will continue to jump forever.)

There is an interesting thing about these jumps. If we mark the points where the frog lands after the first, second, third, and in general, the *n*th jumps, and label the *n*th landing point with the integer n, we can continue numbering forever and never reach the end of the line segment that spans the distance from the frog's starting point to the base of the wall.

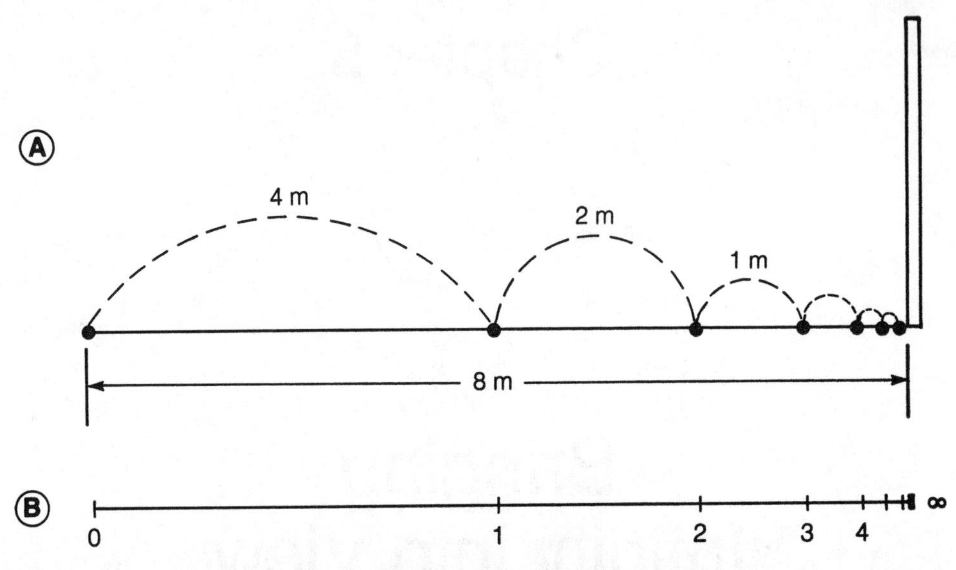

Fig. 5-1. (A) shows the diagram of a hypothetical immortal frog as it jumps halfway to a wall over and over, without end. (B) shows the landing points marked off with positive integers.

That is, we can individually line up all of the positive integers within a finite amount of space, and never are the points crammed so closely together that we cannot distinguish between them. Of course, in the illustration at Fig. 5-1B, the points appear to get squashed together so closely that they cannot be resolved. But that is only because of the limitation of the drawing. If it were possible to actually plot the points, we could keep on doing so without ever getting short of room.

Interpolating between the points, we can see that all of the positive real numbers can fit onto this finite line segment. We can make a number line of finite geometric size and yet have all of the positive reals on it. The number $^1/_2$ would be somewhat closer to the point for the number 1 than to 0; the point for $1^1/_2$ would be closer to 2 than to 1, and so on. We need only guess at these intermediate points, although their exact placement can be calculated. You are welcome to derive the equation for distance versus number value if you like, but it is not especially important for approximate graphs.

We could label the endpoints 0 and infinity (∞); however, we should realize that we haven't really defined the meaning of "infinity." It seems intuitive that the end point on the right of the line segment ought to have this value, but we must define infinity first before we can assign it to this point. We might say, from the drawing, that the following applies.

DEFINITION: Infinity is a numerical entity that has a value larger than the value of any real number. Also, infinity is that numerical entity with the smallest such value.

We know from Cantorian theory of transfinite cardinals that there exists more than one different "size" of infinity. But we will call this "infinity" the smallest one. Is it aleph-null? Probably. Here, we are not so much concerned with that as we are with graphing simple coordinate-plane relations on a quasi-Cartesian system using "squashed" number lines like this.

Two of the number lines from Fig. 5-1B may be placed perpendicular to each other, giving a coordinate system for the first quadrant of the Cartesian xy-plane, which is contained within a finite area. Figure 5-2 is a drawing of this system. The whole xy-plane, including all four quadrants, is shown in Fig. 5-3.

Relations and functions will be distorted on this type of coordinate system. This is inevitable. But for every point on the graph of a relation in regular Cartesian coordinates, there exists one and only one corresponding point on the graph of the relation in these coordinates. The reverse is also true; for every point on the graph of a relation in these coordinates, there is exactly one point in its corresponding graph in regular Cartesian coordinates. There exists a one-to-one correspondence between the points of any relation in either system.

We can employ this "squashing" technique to the radial coordinate of the (r,θ) polar coordinate system also. Then, we have the whole plane inside of a circle. This is shown in Fig. 5-4. We will not be using the polar version of squashed coordinates in this chapter; we will be examining the appearances of relations in the squashed xy-plane. However, you may photocopy the polar system of Fig. 5-4 and graph some relations in the (r,θ) plane if you are interested.

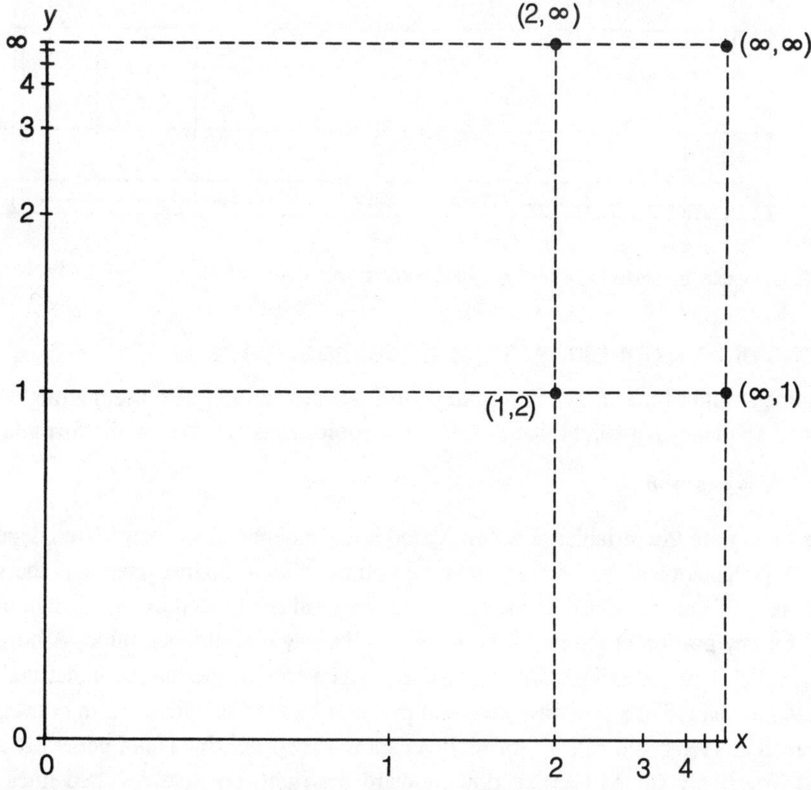

Fig. 5-2. Positive xy-plane (first quadrant) in squashed coordinates, using the frog-and-wall concept for the x and y axes. Some points are illustrated. The points containing "infinity" (∞) are always on the edge of the plane region.

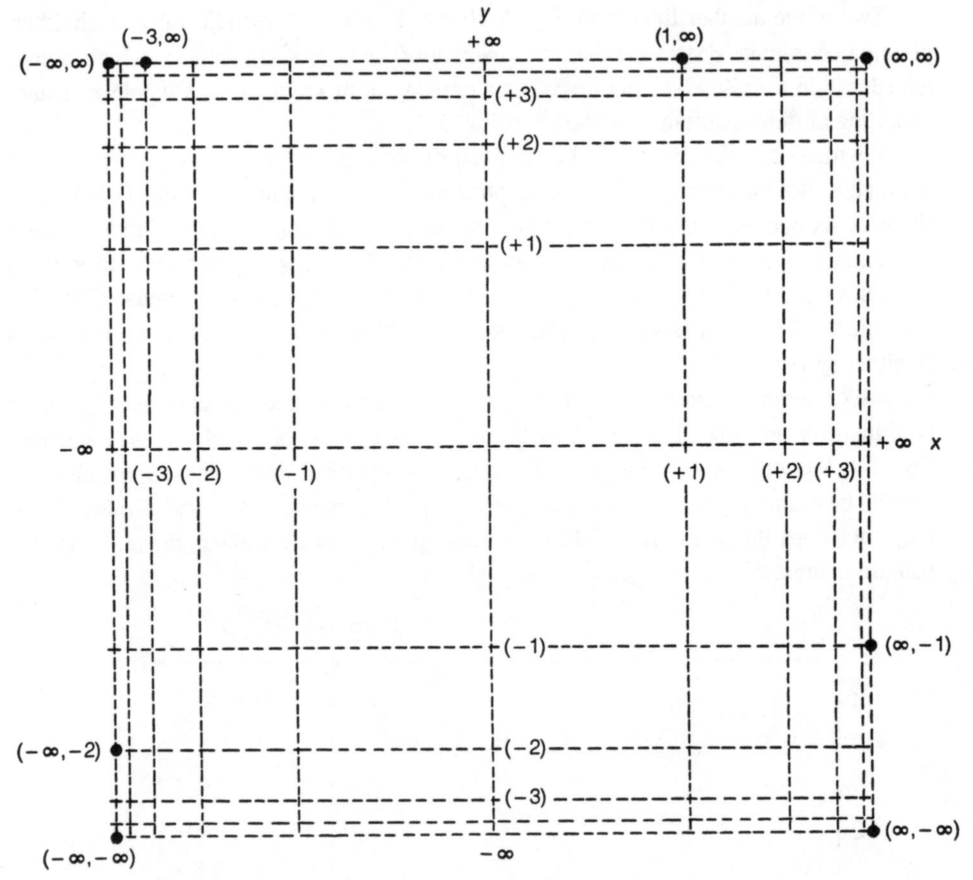

Fig. 5-3. Complete, squashed *xy*-plane, showing some points.

Graphs of Straight Lines in the Squashed *xy*-Plane

Some straight lines in the Cartesian coordinate system will also seem straight in the squashed *xy*-plane. A straight line in Cartesian coordinates is given by the formula

$$y = mx + b$$

where x and y are the variables (x is considered independent and y is considered dependent here), m is the slope of the line, and b is the point at which the line intercepts the y axis. The slope of a line is defined as the change in the y value, divided by the change in the x value, for any portion of the line. The slope may be any real-number value. A horizontal line (parallel to the x axis) has a slope of zero. A vertical line has an undefined slope because the change in x is always zero and division by zero is undefined in conventional mathematics. (We could call it "infinity" in some cases, but this is not generally agreed upon.) Negatively sloped lines go down toward the right; positively sloped lines go up toward the right.

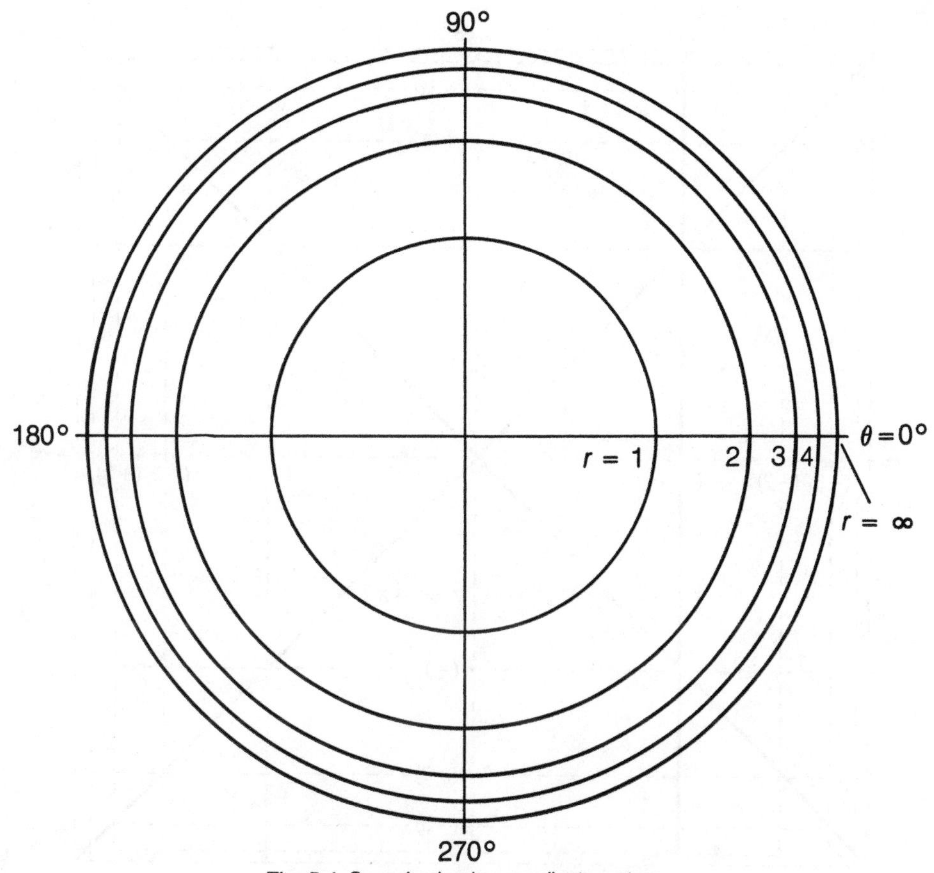

Fig. 5-4. Squashed polar coordinate system.

Any horizontal or vertical line, such as $x = -1$ or $y = 1$, will appear straight on the squashed *xy*-plane. Also, lines with slope 1 or -1 through the origin (0,0) will appear straight; these are the lines $y = x$ and $y = -x$, as shown in Fig. 5-5.

In general, however, lines do not appear straight on the squashed *xy*-plane, even though the same lines appear straight in Cartesian coordinates. Some examples are shown in Fig. 5-6A. The line $y = 2x$, even though it goes through the origin, is distorted because of the more rapid "warping" of the *y* axis relative to the *x* axis. This line reaches greater values of *y* than for *x*, both positively and negatively, so the amount of distortion in the two axes is unequal except at the origin. The distortion is imperceptible near the origin but becomes greater and greater as the distance from the origin increases. The line $y = -(1/3)x$ behaves similarly except that the distortion is greater in the *x* coordinate than in the *y* coordinate. The line $y = x + 3$ does not pass through the origin and is distorted all along its length, either by appearing more vertical, more horizontal, or curved as compared with its rendition in Cartesian coordinates. These same three lines are graphed in Cartesian coordinates in Fig. 5-6B for comparison.

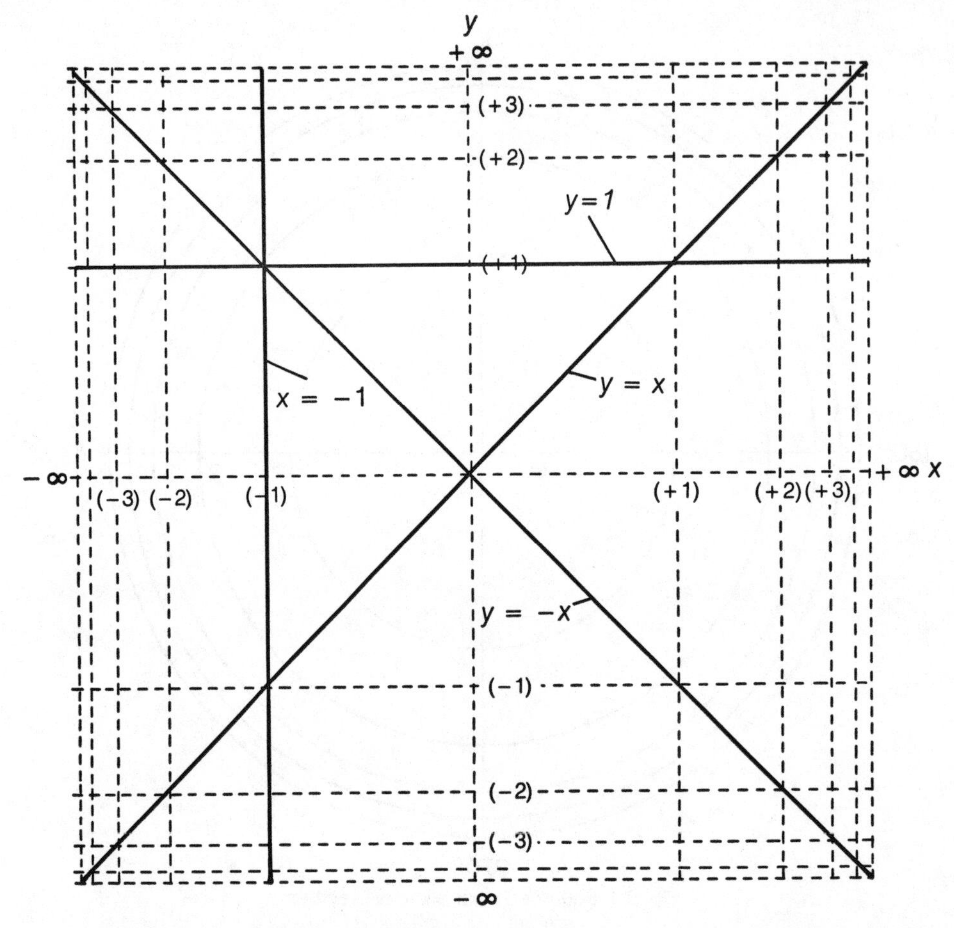

Fig. 5-5. Examples of lines that appear straight on the squashed plane as well as on the Cartesian plane.

Parabolas

Parabolas appear distorted in the squashed *xy*-plane no matter what their formula. In general, a vertical parabola takes the form

$$y = a(x + b)^2 + c$$

where a, b, and c are constants. The constant a determines how "sharp" the parabola will be; the constants b and c determine where the point of symmetry will be.

The parabola $y = x^2$, the simplest, is graphed in squashed coordinates in Fig. 5-7A. Its Cartesian graph is the familiar curve of Fig. 7-5B. Note that the distortion in the squashed plane is symmetrical relative to the y axis. This will be true for any parabola whose point of origin is on the y axis; it will not be true in general for any parabola. An example is the parabola $y = (x + 1)^2 - 1$, shown in Figs. 5-8A and B.

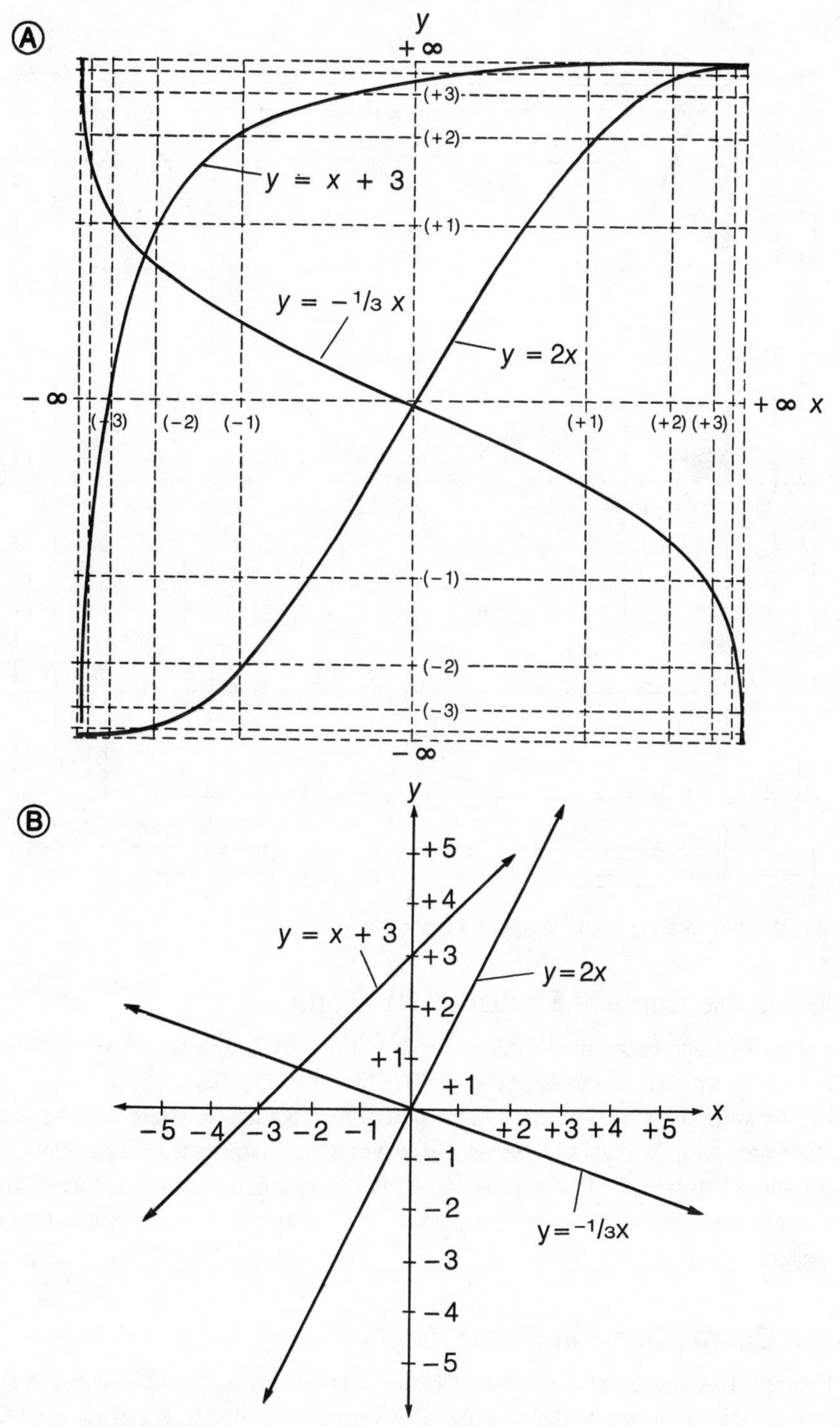

Fig. 5-6. (A) shows some linear graphs that do not appear as straight lines on the squashed plane. (B) shows the same lines in Cartesian coordinates.

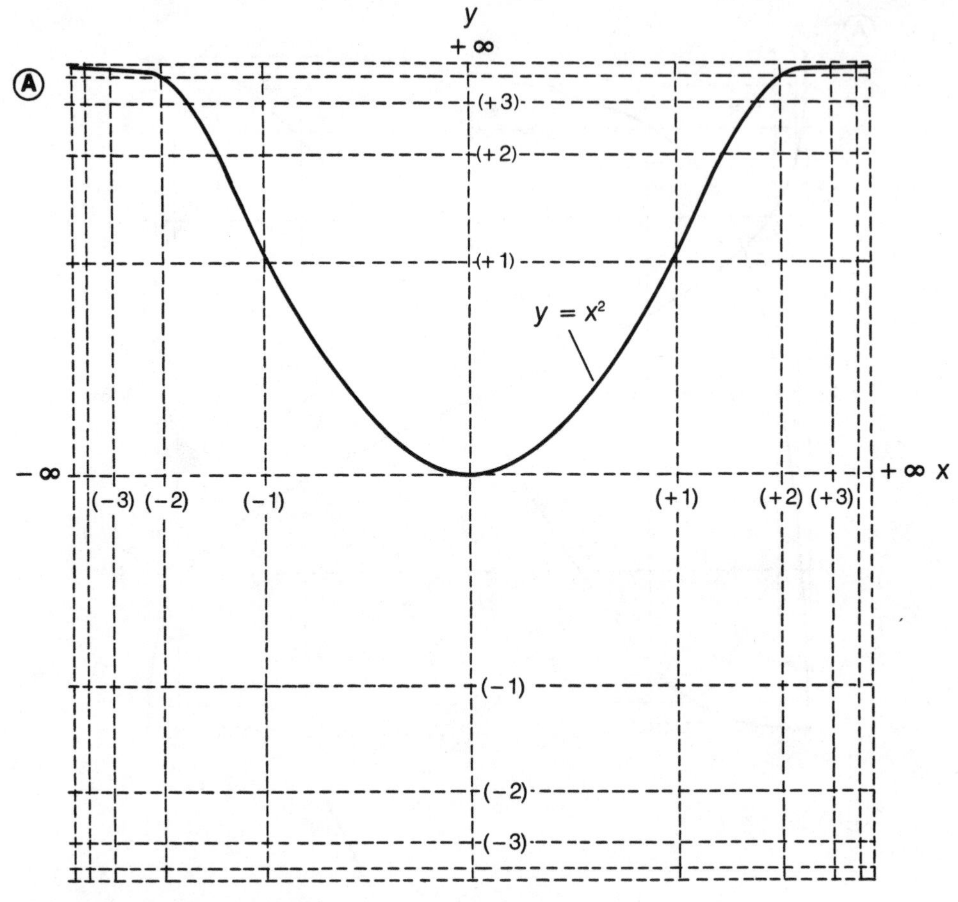

Fig. 5-7. (A) shows parabola $y = x^2$ in squashed coordinates.

Simple Cubic Curve in Squashed *XY*-Plane

Figure 5-9A shows the simple curve $y = x^3$ in the squashed *xy*-plane. The same curve in Cartesian coordinates is shown at Fig. 5-9B. This curve is different from the parabola $y = x^2$ in two ways. First, it is inverted in the part of the coordinate system where x is negative, compared with the parabola. Second, it curves more sharply than the parabola, being closer to the x axis when $-1 < x < 1$, and farther away when $x < -1$ and when $x > 1$. The distortion is more severe also for values $x < -1$ and $x > 1$ in the squashed coordinate system.

Simple Quartic Curve in Squashed *XY*-Plane

Figure 5-10A shows the curve $y = x^4$ in the squashed plane; the same curve in Cartesian coordinates is shown in Fig. 5-10B. This curve is much like the parabola $y = x^2$ except that it is more sharply curved and in fact is even more sharp than the cubic curve.

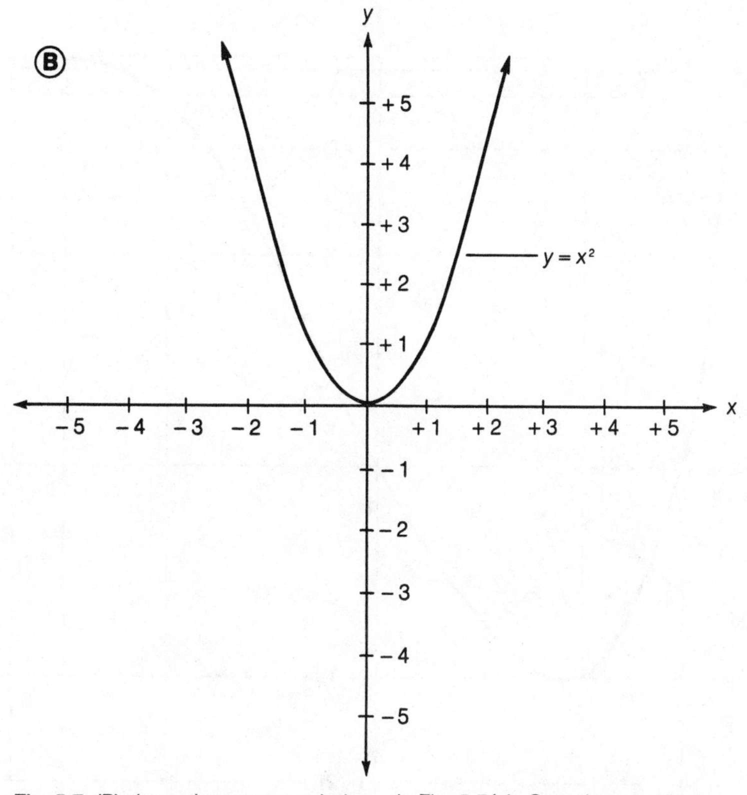

Fig. 5-7. (B) shows the same parabola as in Fig. 5-7A in Cartesian coordinates.

As the exponents increase, the curves alternate between "parabola-like" (where y is never negative) and "cubic-like" (where the range of y values is the entire set of reals, and the left-hand part of the curve is inverted). However, they get sharper and sharper as the exponents increase, approaching squared-cornered curves. They are never quite perfectly square for any finite value of exponent.

Extreme Exponent Values

The curve for the equation $y = x^{1000}$ is shown in Fig. 5-11 in squashed coordinates. Note that the corners appear perfectly squared off. Actually they are not; the curve is still a "curve." If we were to look very closely at it, for example with a high-powered microscope, and if we could make the lines thin enough, this would become apparent.

The curve for $y = x^{1001}$ is at Fig. 5-12. This curve resembles the cubic curve rather than the parabola in that the left-hand side (where $x < 0$) is in the third quadrant; $y < 0$. This curve also appears squared off and, in fact, is a little more so than the curve for $y = x^{1000}$. But it, too, is still slightly rounded off. No matter how large we make the exponent, the curve will always be at least slightly curved, never perfectly square. However, it will

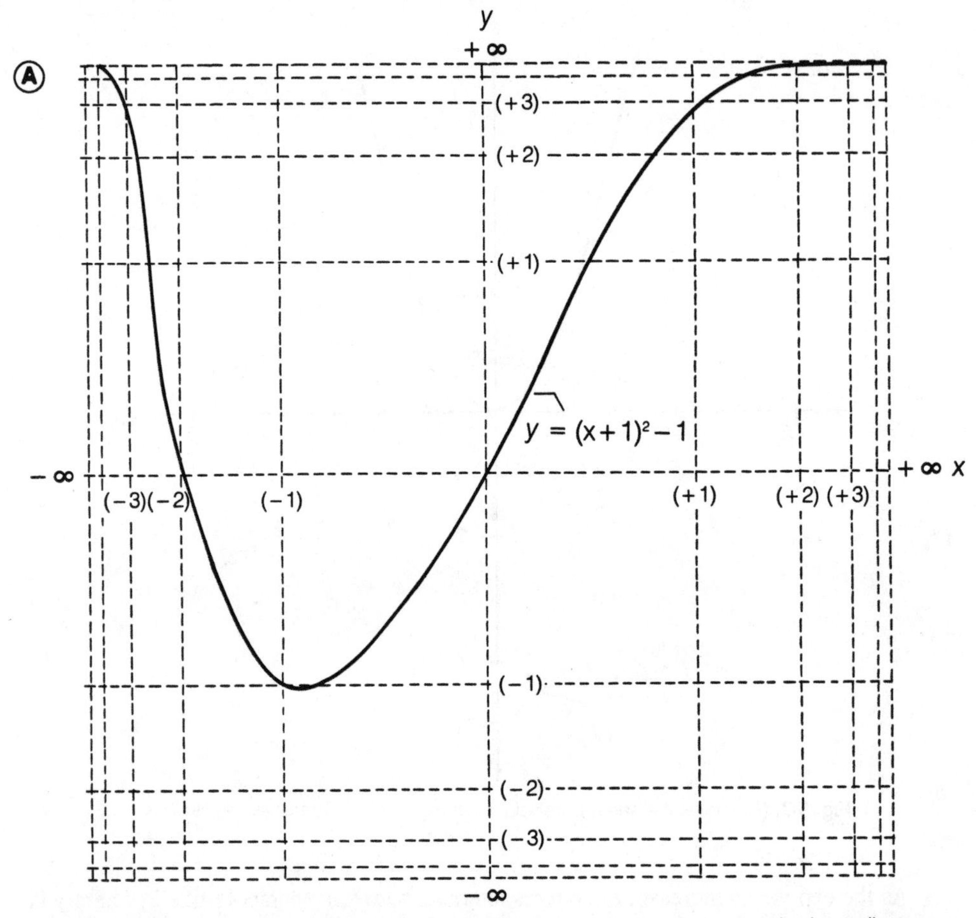

Fig. 5-8. (A) shows an example of a parabola that is distorted asymmetrically in squashed coordinates: $y = (x + 1)^2 - 1$.

continue to alternate as the integer increases, with odd exponents resulting always in inverted left-hand parts of the curves.

We might speculate what would happen if we could raise the value x to the infinitieth power. Cantor in fact did this with his transfinite cardinals, raising two to the power aleph-null and proving that the result was aleph-one, or at least, that the result was larger than aleph-null. In our case here with our system of graphs, we will not worry about the transfinite cardinal value of the "infinity" that occupies the edge of the plane region. We simply use the symbol for infinity (∞), and we do not make a distinction as to whether it might be aleph-null or aleph-one or whatever.

We could say that the graph for $y = x^\infty$ would indeed have perfectly squared-off corners. The value for $x = 1$ would be multiple, ranging all through the nonnegative reals, including infinity as well. The value for $x = -1$ would range through all the reals, and positive and negative infinity also. For $-1 < x < 1$, $y = 0$. For $x < -1$, $y = \pm\infty$. For $x > 1$, $y = \pm\infty$. The graph is illustrated in Fig. 5-13A in squashed coordinates.

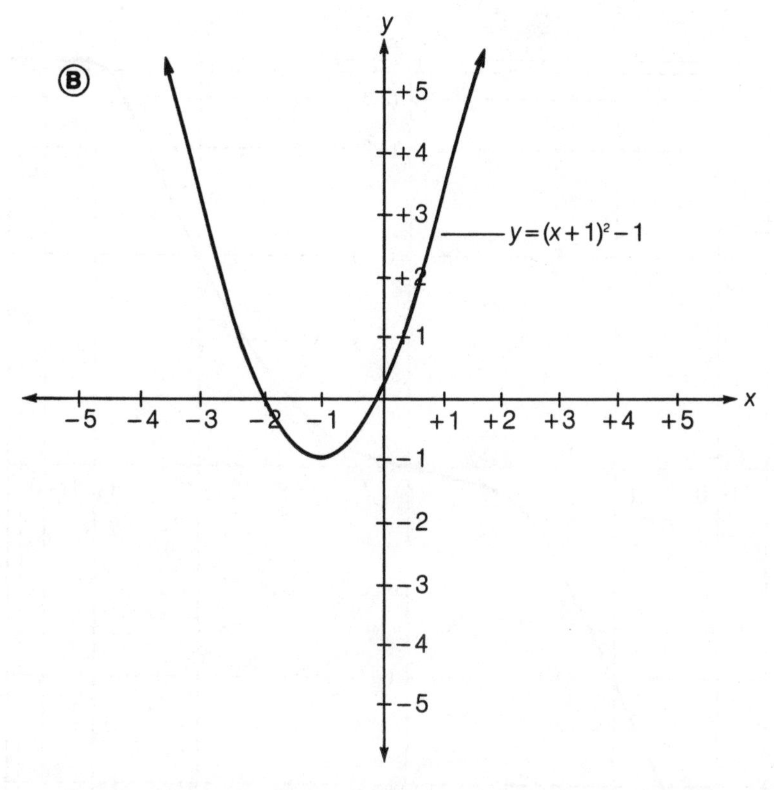

Fig. 5-8. (B) shows the same parabola graphed in Cartesian coordinates.

This involves a certain amount of speculation. For a number between 0 and 1, for example, the value approaches zero as the exponent grows without bound. But it is a stretch of the imagination to say that raising this number to the *infinitieth* power will actually yield *zero*. We have not proven this. Although it seems intuitively plausible, it must be proven in order to be established as a mathematical fact. Thus, in a sense, since we will not try to prove it here, we are playing a game—nothing more. This is not offered as any great new mathematical theory.

Similarly, we have not proven that 1^{∞} is the whole set of nonnegative reals, along with $+\infty$. Nor have we proven any of the "facts" that give us the graph in Fig. 5-13A. What we have done is observed the way the graphs behave as the exponent grows without bound and then simply defined the graph for an infinite exponent as the limiting shape of the other graphs. In so doing, we say that infinity could be both even and odd, rather than saying that it is neither even nor odd.

The limiting shapes are in agreement between squashed and Cartesian coordinates, although in Cartesian coordinates, we do not see the entire graph of the equation $y = x^{\infty}$. Figure 5-13B shows the graph in Cartesian form; the values for $x < -1$ and for $x > 1$ are not visible because they are infinite. These values *are* quite visible in the squashed system as depicted in Fig. 5-13A.

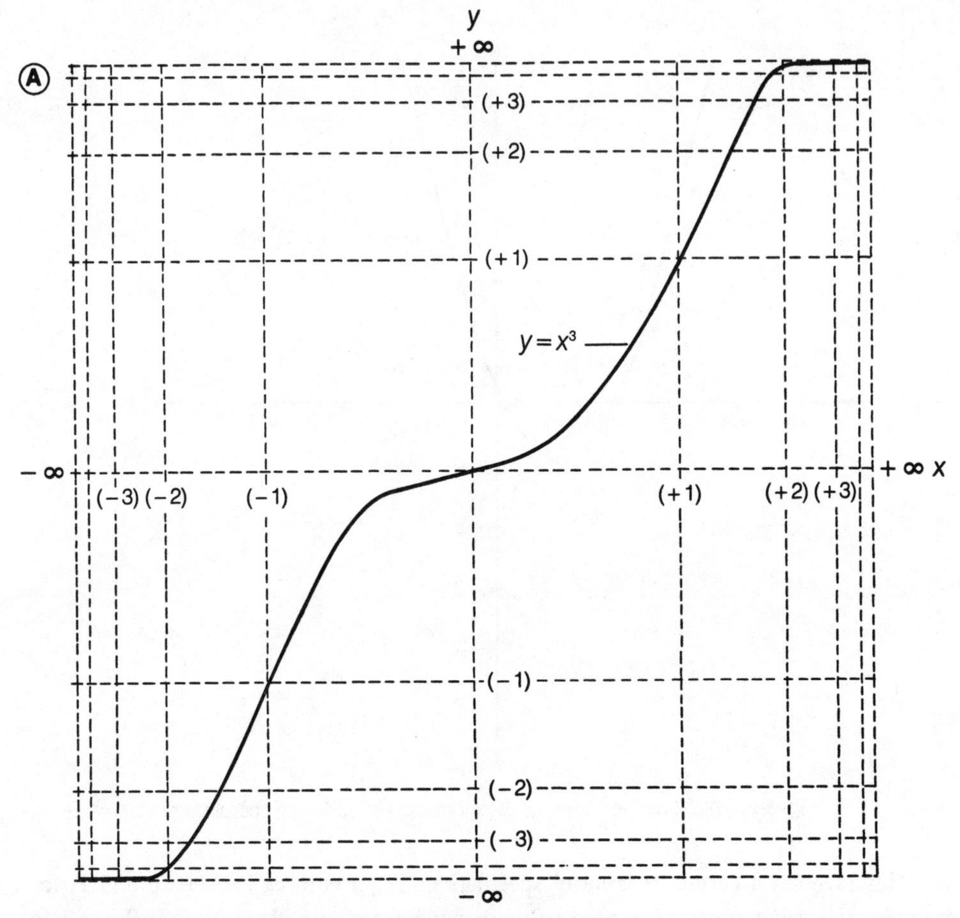

Fig. 5-9. (A) shows the cubic curve $y = x^3$ in squashed coordinates.

Larger and Larger Circles

The graphs of circles in the squashed *xy*-plane are also quite interesting. The general form for the graph of a circle is

$$(x - a)^2 + (y - b)^2 = r^2$$

where *a*, *b*, and *r* are constants. The radius of the circle is *r* units on the coordinate plane. The values *a* and *b* determine the center of the circle. When $a = 0$ and $b = 0$, the center of the circle is the origin (0,0).

In Fig. 5-14, circles at the origin with radii of 1, 2, and 4 units are shown. At (A), the circles are graphed in squashed coordinates. In (B), they are graphed in Cartesian coordinates. Note that the graphs in squashed coordinates are distorted, resembling television picture tubes (circa, perhaps, the mid 1950s). The circles become more and more nearly square as the radius increases. This distortion is caused by the nonlinearity of the two axes.

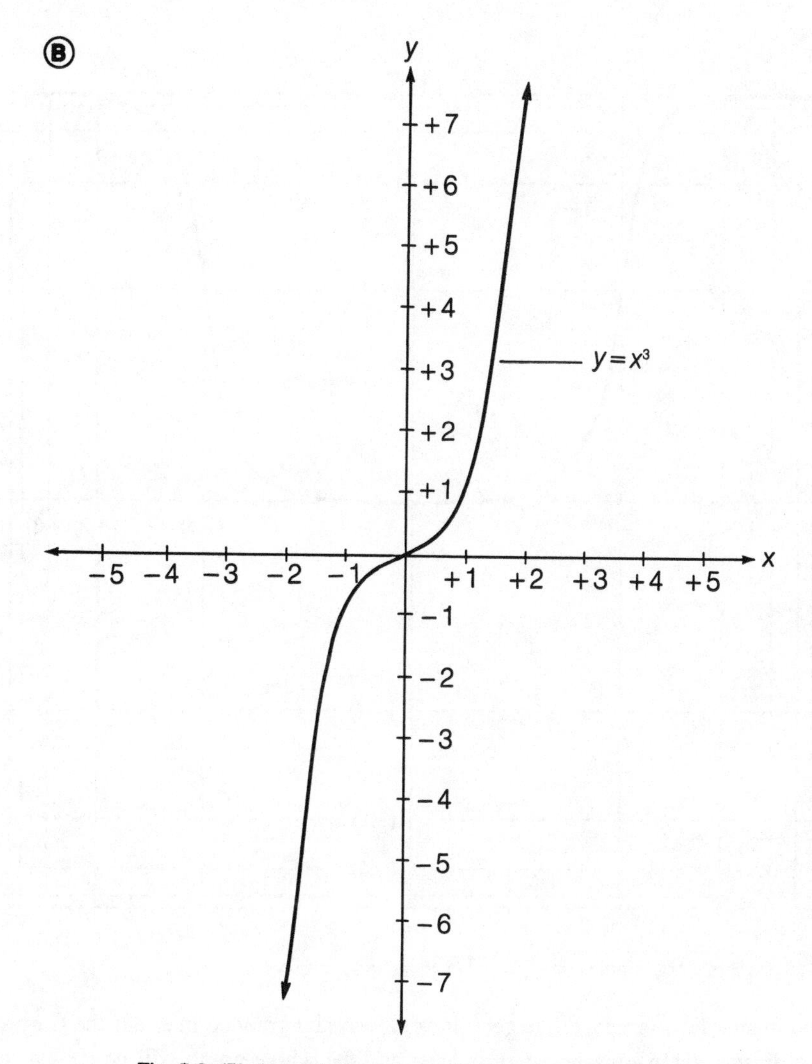

Fig. 5-9. (B) shows the same curve in Cartesian coordinates.

Circles with centers not at the origin will have asymmetrical shape in squashed coordinates, just as parabolas and other exponential curves are unevenly distorted when they are not symmetrical with at least one of the coordinate axes. An example is the circle $(x - 1)^2 + (y - 2)^2 = 4$, shown in squashed coordinates at Fig. 5-15A and in Cartesian coordinates at B.

When the radius becomes very large, the circle appears to occupy the peripheral square of the squashed plane. The axes could be scaled up, for example, by numbering them 10, 20, 30, and so on rather than 1, 2, 3, and so on. However, for any scheme of numbering the axes, there are circles so large that they will appear square as graphed in the squashed coordinate plane. An example is shown in Fig. 5-16 for $x^2 + y^2 = 10{,}000$.

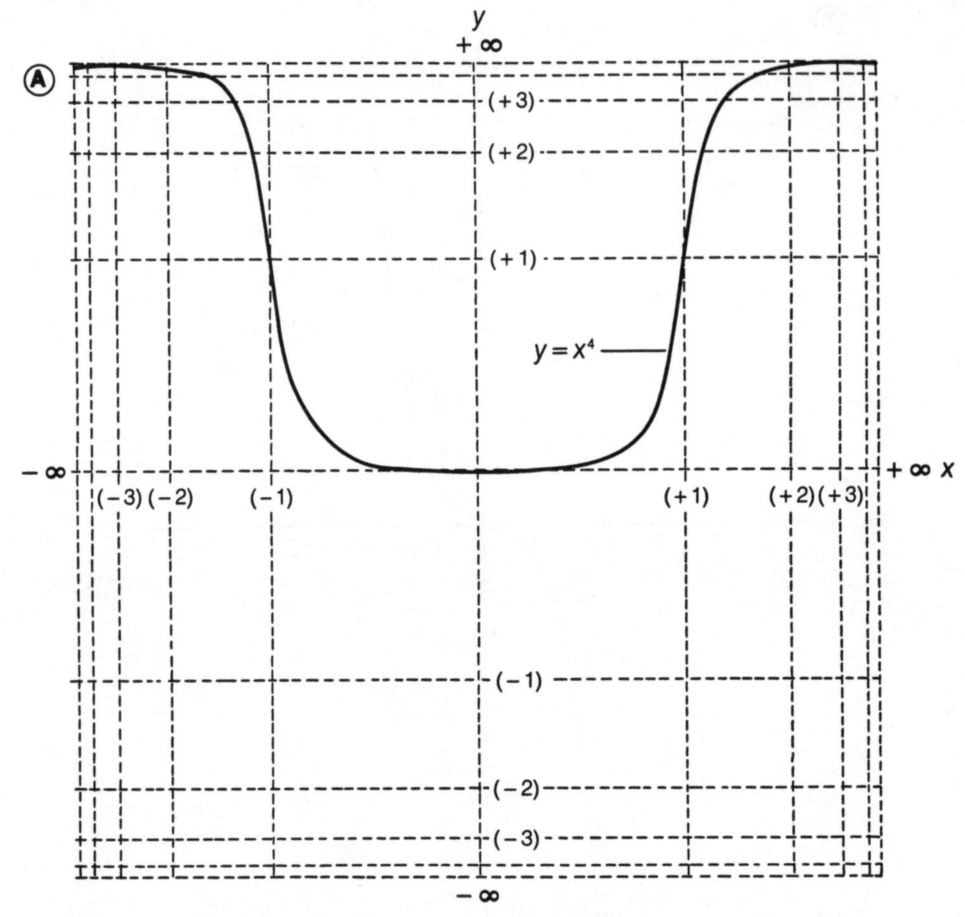

Fig. 5-10. (A) shows the quartic curve $y = x^4$ in squashed coordinates.

No matter how large a circle gets, it can always be graphed in either the Cartesian set of coordinates, or in squashed coordinates, and the whole graph will be shown, not just part of it, as is the case with the parabola in Cartesian coordinates. But suppose we carry the equation for the circle to its extreme limit so that it actually becomes the square bordering the squashed coordinate system. Then we might say that the radius is infinite and, since $\infty^2 = \infty$, the equation is

$$x^2 + y^2 = \infty$$

We say that $\infty^2 = \infty$ because it cannot be finite. Such a "circle" will not show up at all in Cartesian coordinates, no matter how large we make the graduations on the axes.

What Are We Really Looking At?

When we squash up the coordinate system as we have done in this chapter, what are we actually doing? Are we really looking at the infinite by means of some mathematical/

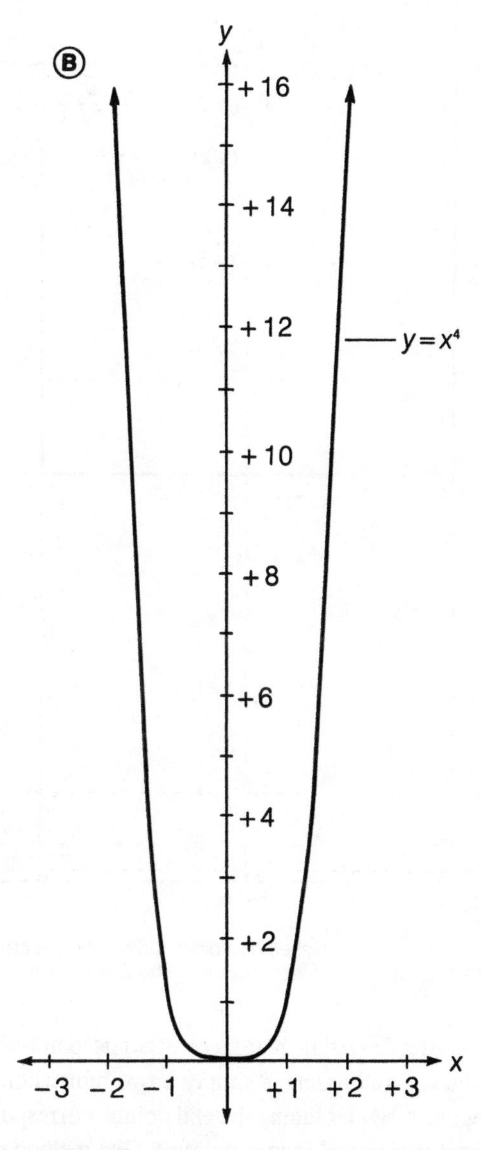

Fig. 5-10. (B) shows the same curve in Cartesian coordinates.

geometrical magic? We cannot say. We haven't done any rigorous mathematical derivations here. It is intuitively appealing to say that we are looking at infinity.

We are, to be sure, graphing the entire set of points for functions such as the parabola or cubic or quartic curves. These are functions that cannot be shown in their entirety on the Cartesian plane because they are infinite curves. We can prove that each point on the squashed plane corresponds to one unique point in the Cartesian system, and vice versa, if we consider the square border of the squashed system *not* to be a part of the actual system of coordinates.

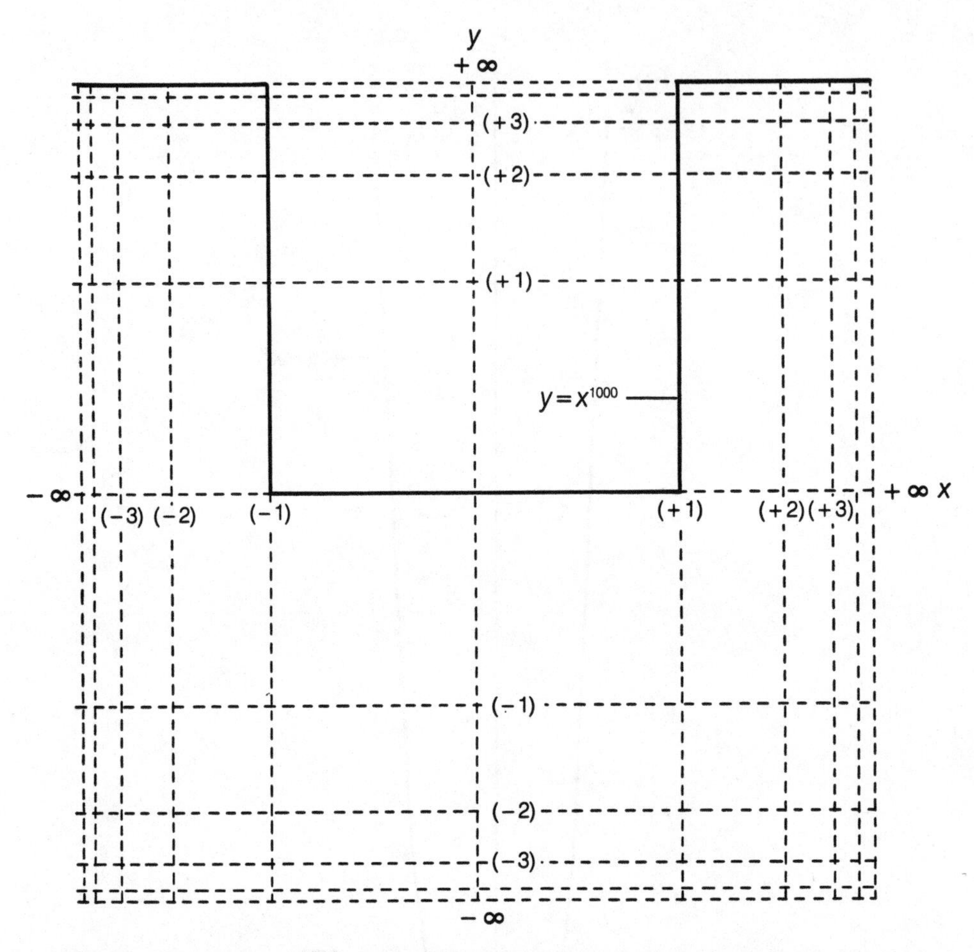

Fig. 5-11. The curve for $y = x^{1000}$ in squashed coordinates. The corners are not truly square, but they appear that way in the drawing because of their extremely sharp curvature.

Proving that the entire Cartesian plane corresponds point-wise with the squashed plane (not including the square border) is simply a two-dimensional case of proving that the points on a line segment, not including the end points, correspond one-to-one with the real numbers. This proof is classical in mathematics. One method of geometrically showing this proof is to consider some line segment *S* of length *L*. We bend this line segment into a half-circle without altering its length. At the center point of the half-circle is the radial point we shall call *P*. We draw a line segment *T* from *P* through the exact middle of semicircle *S* and extend *T* down past *S* somewhat where it intersects a base line *M*. The line *M* is perpendicular to *T*. All of this is shown in Fig. 5-17.

Let the line *M* represent the set of real numbers. Presume that the set *R* of real numbers is in one-to-one correspondence with the points on *M*. Now we may draw a radial line *Q*, from *P* through *S*, and it will always intersect *M* at some point. Let the points be called *QS* and *QM*, after the lines that intersect to form them. We assign each point *QS* on *S* the

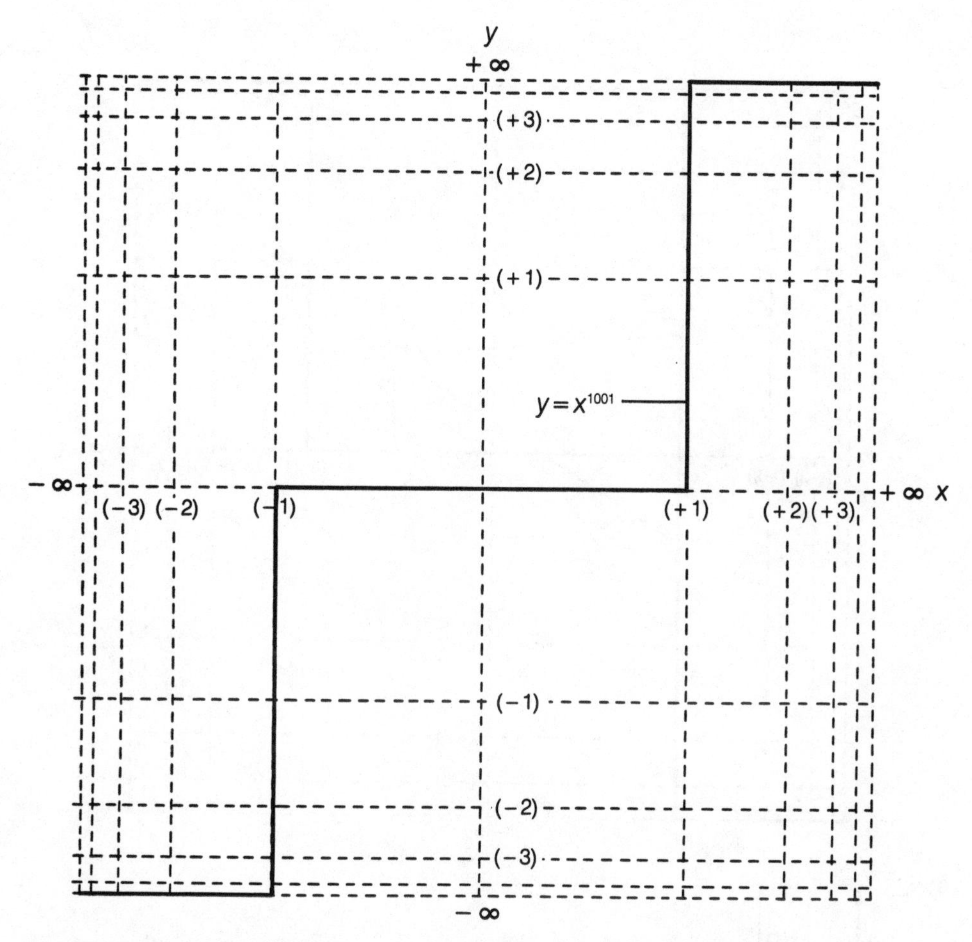

Fig. 5-12. The curve $y = x^{1001}$ in squashed coordinates. This curve is almost identical to the one in Fig. 5-11 except that the negative x part is inverted. Actually, the corner curves are just a little bit sharper than is the case for the curve $y = x^{1000}$.

real number that corresponds to QM on M. Clearly, every point on M will have a point on S that can be formed in this way. Every point on S, except for the actual end points, will have one and only one correspondent on M. Thus, the points on S can be assigned the real numbers. In fact, the squashing of the real numbers on S that occurs in this example is similar to that we have employed in the squashed coordinate system in this chapter. The actual function is a little different, but the effect is exactly the same.

In the proof that the Cartesian plane and the squashed plane have a one-to-one correspondence of points, we consider the ordered pairs (x,y). If the points in the squashed plane are called (x_s,y_s) and those in the Cartesian plane are (x_C,y_C),then we know from the geometric example of Fig. 5-17 that there is a one-to-one correspondence between the x_s and x_C. In the same manner, there exists a one-to-one correspondence between y_s and y_C. In fact, the transformation functions are exactly the same if the axes of the two systems are

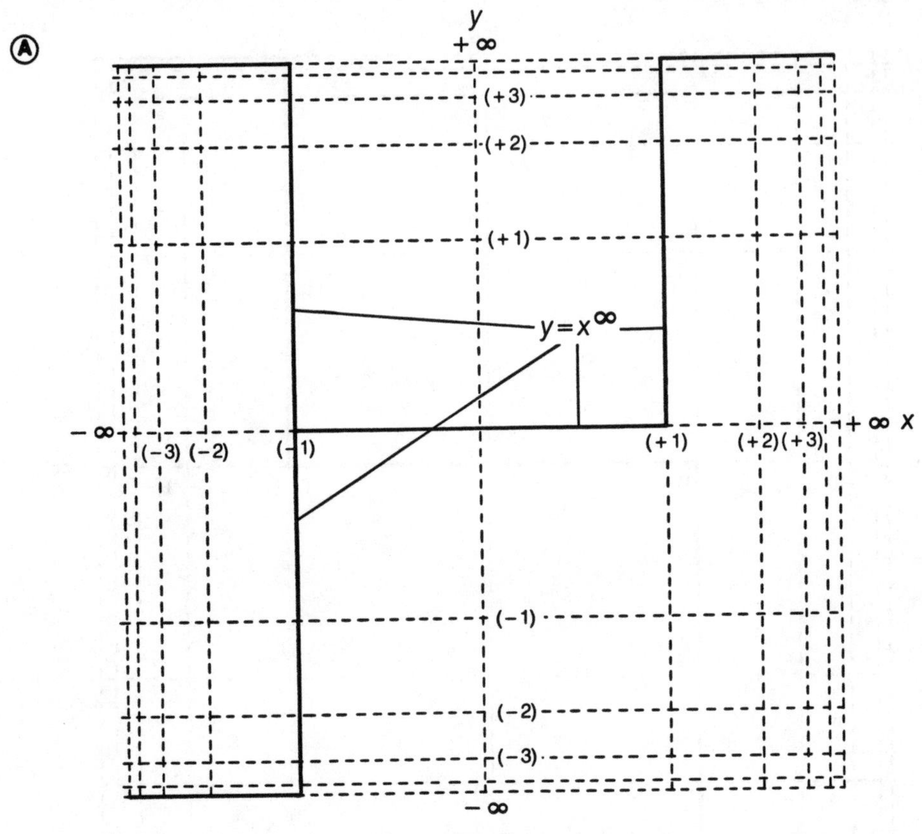

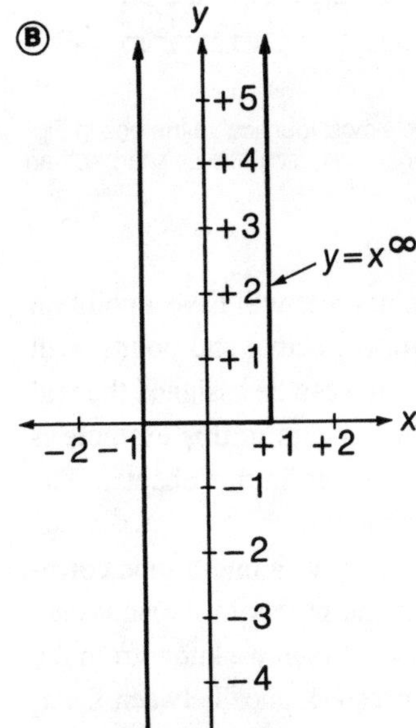

Fig. 5-13. Hypothetical graph of $y = x^{\infty}$, based on the notion that ∞ could be both even and odd. These corners actually are perfectly square; at (A), in squashed coordinates, and at (B), in Cartesian coordinates. Part of the graph cannot be depicted in the Cartesian plane no matter how large we make the values on the axes.

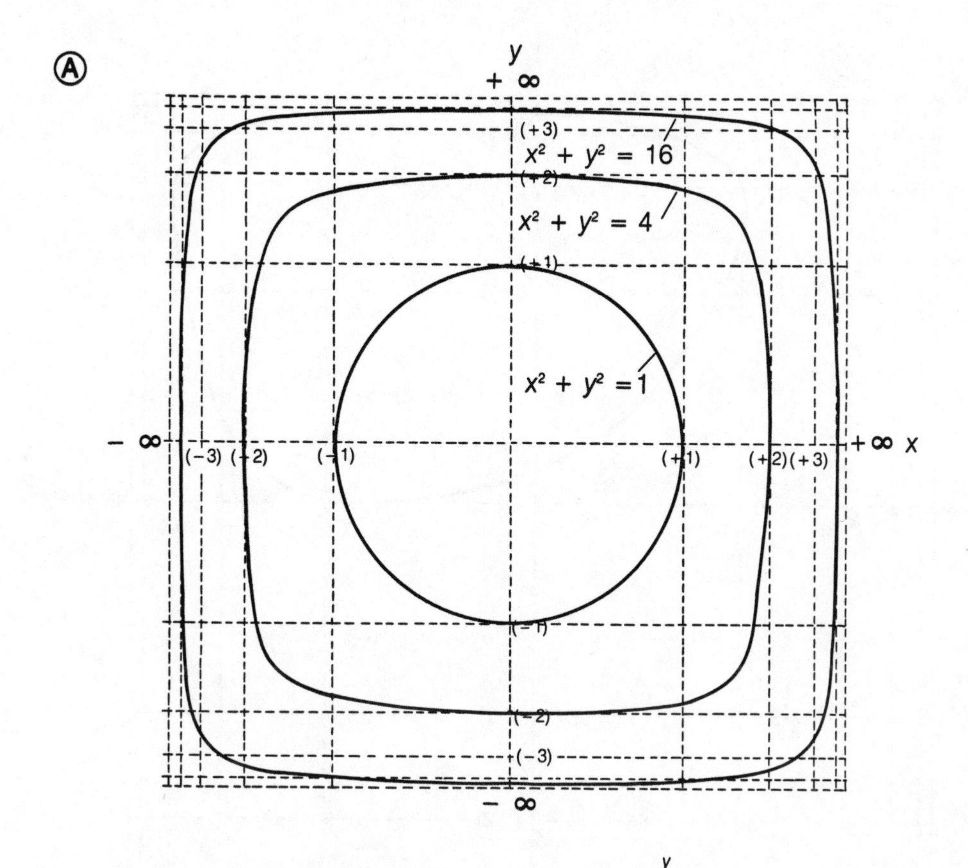

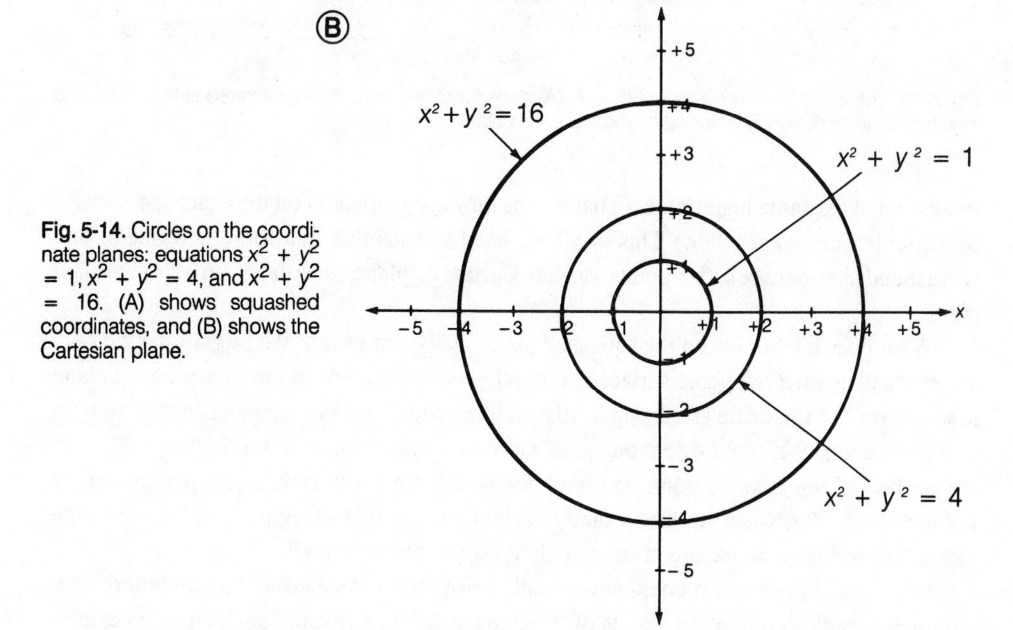

Fig. 5-14. Circles on the coordinate planes: equations $x^2 + y^2 = 1$, $x^2 + y^2 = 4$, and $x^2 + y^2 = 16$. (A) shows squashed coordinates, and (B) shows the Cartesian plane.

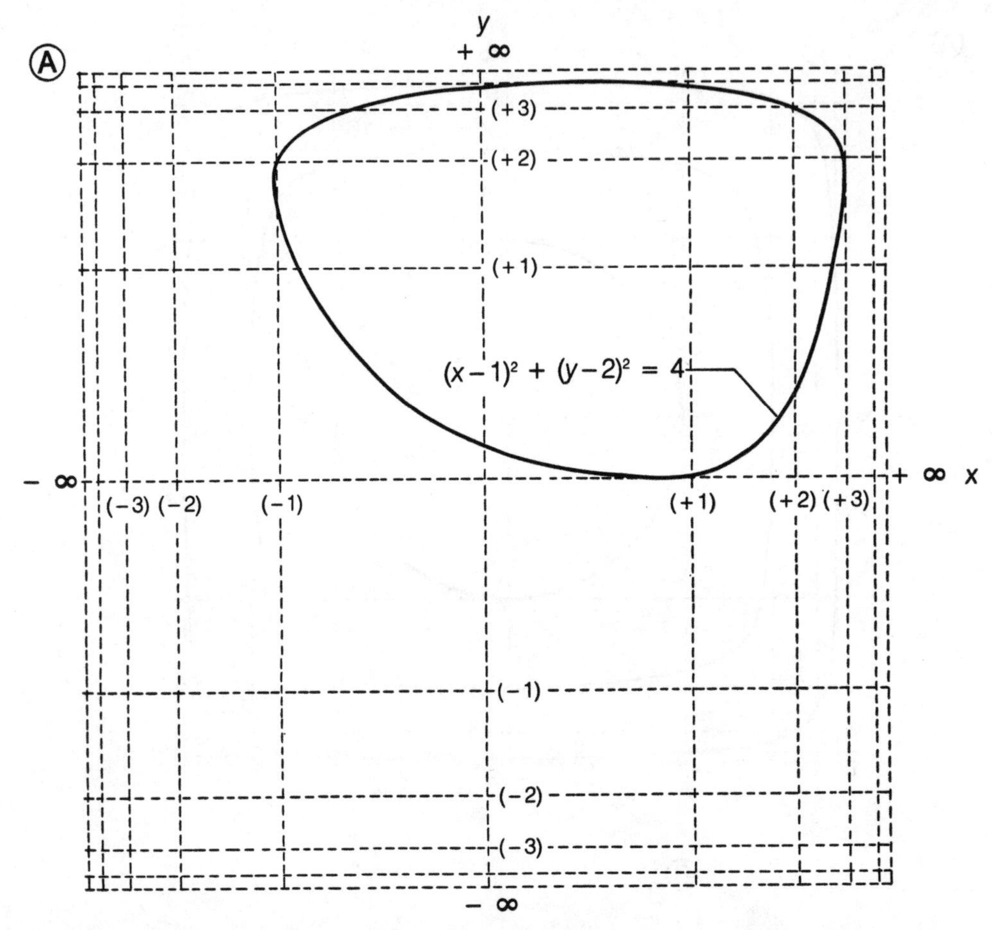

Fig. 5-15. The circle $(x - 1)^2 + (y - 2)^2 = 4$. (A) shows that this circle is not centered at the origin and therefore is asymmetrically distorted on the squashed plane.

numbered in the same magnitude. (That means they are graduated on the exact same scale, being replicas of each other.) This is all we need to establish that there is a one-to-one correspondence between the points on the Cartesian plane and those on the squashed plane.

What does the border on the squashed plane really represent? We might imagine ourselves standing on a flat plane surface and looking outward at the plane. Suppose the plane is scaled off like quadrille graph paper, like an immense tiled floor going infinitely away in every direction. We could direct our gaze more and more nearly to the horizon. We will always be looking at some point on the plane, perhaps a few meters away, perhaps many kilometers off. But when we look exactly horizontal, our line of sight never intersects the plane. We will then be gazing at infinity. Is it on the plane or not?

Of course this situation could never really arise, and physicists and astronomers have arguments ready to refute our claims of "looking at infinity" even if such a plane actually

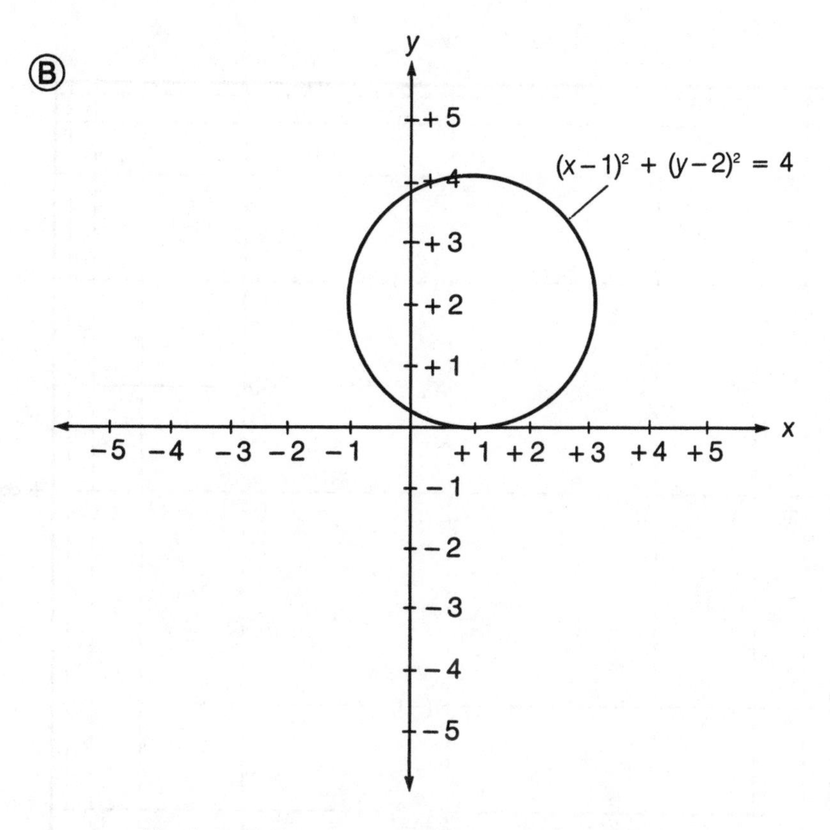

Fig. 5-15. (B) The graphed Cartesian coordinates of Fig. 5-15A.

is found. Our field of sight, as we stand on such a plane, is squashed because the entire field of our sight at any given time can be represented by a celestial sphere. The plane is, in fact, visually squashed exactly as the number line is compressed into the half-circle in Fig. 5-17.

This is all based on Euclidean geometry. But it is believed that the real universe is non-Euclidean. Thus, we have to throw out the very notion of a geometric concept of infinity, at least in this universe.

Where does this leave us? Playing games for nothing, about nothing? Perhaps. But for every finite quantity, there exists something more after that. Even in a finite-but-unbounded, non-Euclidean universe, there must be the implied existence of another dimension in which it is "bent," and in this dimension, who can say what we might find?

The behavior of the infinity we discover by squashing the Cartesian plane can be examined, and we find that:

- There exist an infinite number of points at infinity;
- There are positive and negative infinities and they are not the same;
- There must be something outside the squashed plane that cannot be defined even according to our definition of infinity as the border of that squashed plane.

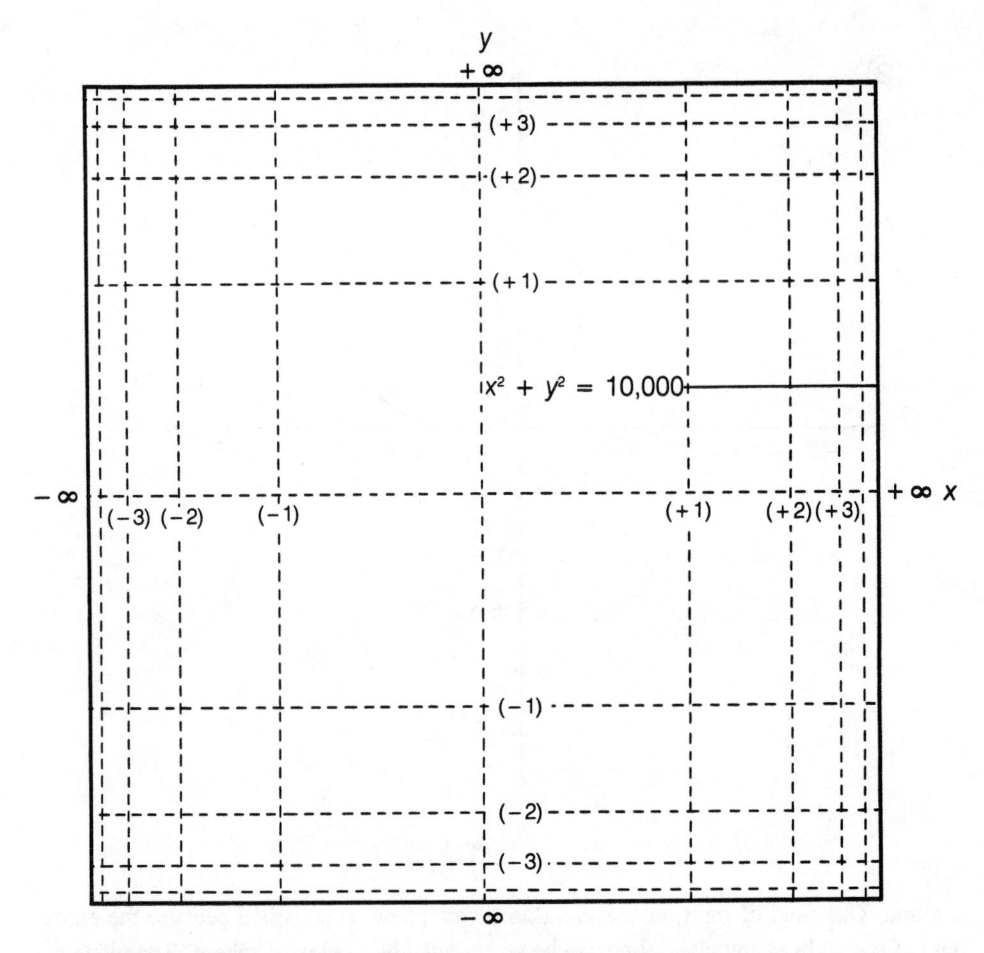

Fig. 5-16. The graph of the large circle $x^2 + y^2 = 10{,}000$ in squashed coordinates. The radius of this circle is 100. It looks like a square at the boundary of the squashed plane; actually its corners are curved and it is just inside the boundary. The graph of an infinitely large circle would indeed occupy the boundary of the squashed plane.

What is that? It is what we see when we look up above the horizon into the heavens. No matter how much we define things as we see them, there is always more that we have yet to see.

Perhaps there is some accommodation outside the squashed plane for Cantor's hierarchy of transfinite cardinals. Perhaps that space can only be filled by postulating that there are infinitely many universes. The imagination is the limit.

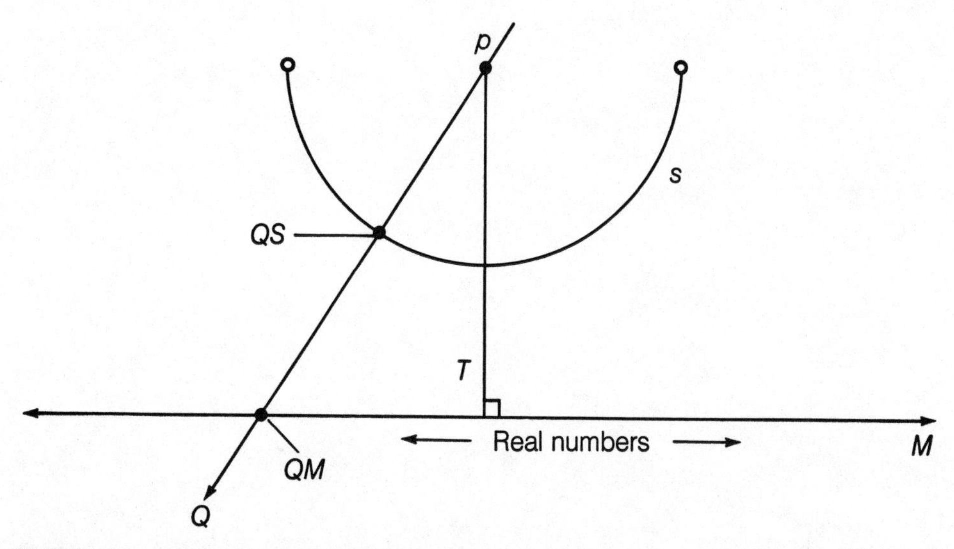

Fig. 5-17. Geometric illustration of one-to-one correspondence between the points on line segment *S* and a line *M*. (Details are in the text.)

Chapter 6

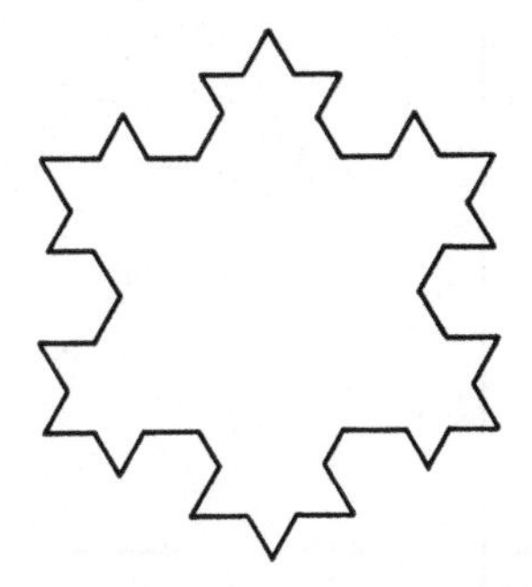

Geometric Infinities

WE HAVE APPROACHED INFINITY FROM THE STANDPOINT OF SET THEORY, FROM THE standpoint of algebra, and, in the last chapter, from the point of view of analytic geometry (the theory of graphs). We have been, of necessity, quite informal and nonrigorous with this, mainly to keep things from getting cumbersome, dry, and complicated. That approach continues here. In fact, we will be loosening things up even further. I run the risk of offending mathematical purists this way, perhaps, and if this seems to be a silly game, let me say with enthusiasm: "Let the game go on!"

The concept of infinity appears in certain extreme geometric cases. In high-school geometry, when mathematical proof is usually introduced for the first time in its full form, some cases arise that we simply dismiss as "undefined." There is no such thing, for example, as a 0-90-90 triangle. The sum of the measures of the interior angles of a triangle is always 180 degrees in Euclidean geometry, but none of these angles may measure zero, 180 degrees, or anything more than 180 degrees. So while we may have a 30-60-90 triangle, or a 45-65-70 triangle, we can't have 0-90-90 or 0-0-180 "triangles." These figures wouldn't be triangles. They might look like the figures in Fig. 6-1A and B.

But the 0-90-90 and 0-0-180 "triangles" can be regarded as infinitely tall and infinitely short, respectively. Why can't triangles have infinite dimensions? If we refuse to define such things, we must ask why, and then we ought to at least make an attempt to provide some answer to this question. Usually, the answer takes the form of a certain kind of facial expression suggesting disdain or perhaps pity and almost invariably disgust, followed by a reversion to some other topic or a disguised insult.

That doesn't do anything to solve the problem at hand.

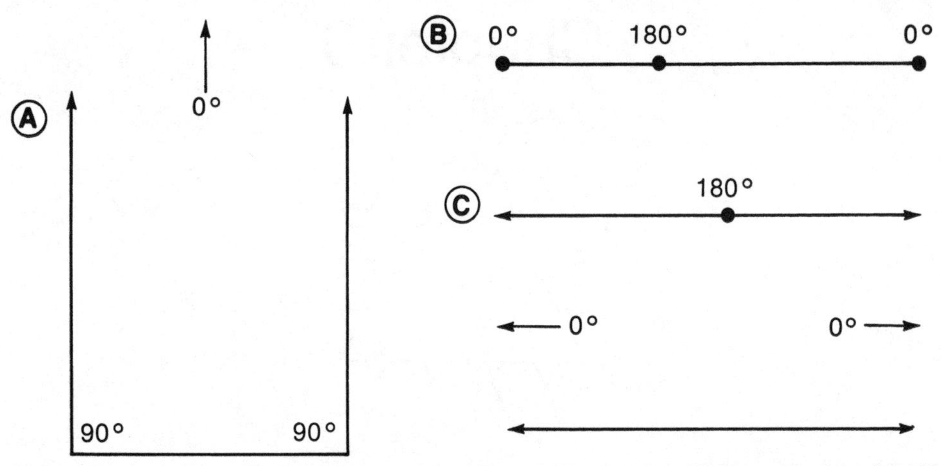

Fig. 6-1. Extreme triangles. (A) shows 0-90-90 and (B) and (C) show two different types of the 0-0-180 triangle.

The Concept of Infinite Space

We regard Euclidean space as infinite by its very nature. So we ought to have room, in theory at least, for infinitely tall objects such as the 0-90-90 triangle. We might have another form of 0-0-180 triangle, such as is illustrated in Fig. 6-1C. It looks like two parallel lines, with the 180-degree angle within sight on one of the lines and the two 0-degree angles infinitely far away in opposite directions.

Who is to say what happens "at infinity" in an infinite space? Perhaps we can escape this whole question by means of the now-accepted theory that space, in reality, is finite and yet unbounded, the three-dimensional surface of a four-sphere. Then there really cannot be parallel lines that never meet; there cannot be "triangles" such as those in Fig. 6-1A and C. Yet in the mindscape, Euclidean, infinite space *is* possible. So the question of what parallel lines "do" when the "reach infinity" is a valid question. They may remain parallel and at the same distance; they may intersect, or they may even diverge. How can we tell?

We might use the infinite coordinate (IC) plane from the last chapter to attempt to answer this question. We can construct parallel lines, say

$$L_1: y = x + 1$$
$$L_2: y = x - 1$$

These are lines with slope 1 and intersecting the y axis at 1 and -1. The renditions on the IC plane and the Cartesian plane are depicted in Fig. 6-2A and B.

While these lines are parallel in the sense that they are equidistant at all points on the Cartesian plane at (B), they seem to converge at the positive and negative infinities on the IC plane.

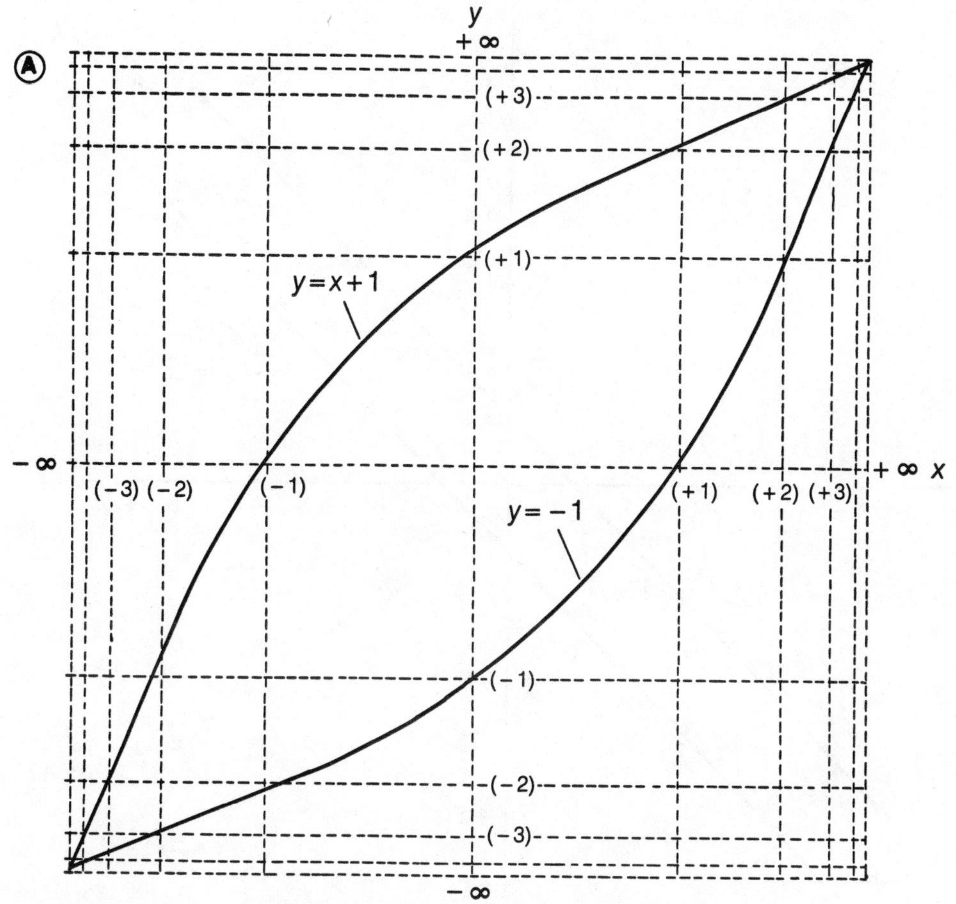

Fig. 6-2. When graphed on the infinite coordinate (IC) plane, the parallel lines $y = x + 1$ and $y = x - 1$ appear to intersect at the positive and negative infinite extremes, as shown at (A).

Let us now investigate the lines

$$L_3: y = 1$$
$$L_4: y = -1$$

These lines are also parallel, and they seem to maintain their distance from each other even at infinity as seen on the IC plane (Fig. 6-3A). Of course they appear parallel on the Cartesian plane (Fig. 6-3B).

No matter what kinds of examples we use for parallel lines, it is apparent that they will either converge, as in Fig. 6-2, or remain at the same distance, as in Fig. 6-3. They will never diverge if they are parallel. We cannot find an orientation for two parallel lines on the IC plane where they will appear to diverge at infinity. As a matter of fact, the only cases where parallel lines will remain equidistant and not converge are those special situations where the slope is either zero (the lines are horizontal) or not defined (the lines are

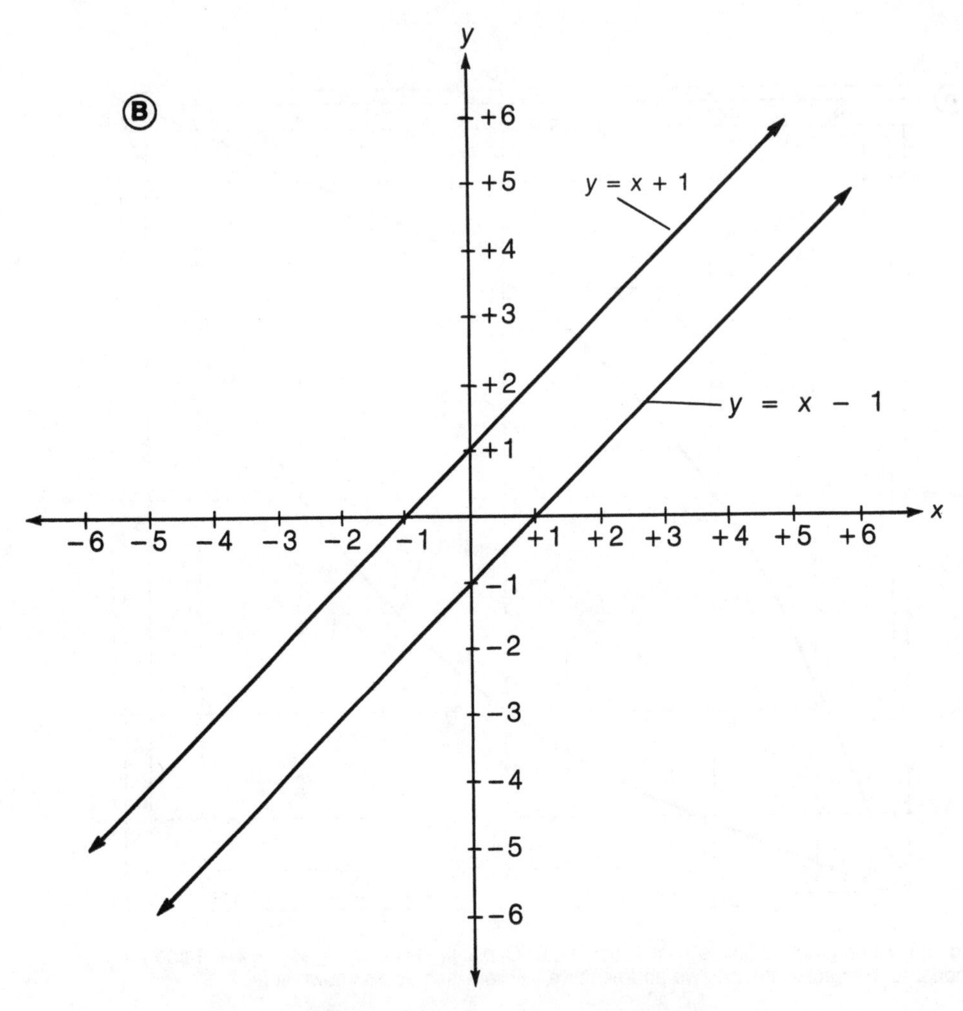

Fig. 6-2. (B) shows the lines are graphed on the Cartesian system.

vertical). Any two parallel lines, where the first is chosen at random, will converge at infinity in either direction: such "random" lines would *never* be exactly horizontal or exactly vertical. There are infinitely many ways a randomly chosen line might be sloped and only two ways in which the line could be exactly horizontal or vertical.

Is this a proof that parallel lines in Euclidean space converge at infinity? Not a rigorous proof. But it suggests that we are all right to suppose that they do. Then we can indeed have extreme triangles such as the ones depicted in Fig. 6-1A and Fig. 6-1C.

Just as an infinite line is presumed to be in a one-to-one correspondence with the real numbers, or at least some facsimile of the line is in such correspondence, an infinite plane is in one-to-one correspondence with pairs of real numbers (x,y). The infinite three-space can be paired off pointwise with ordered triples (x,y,z) of real numbers. There is some question as to whether the real numbers can account for every point on a line or whether

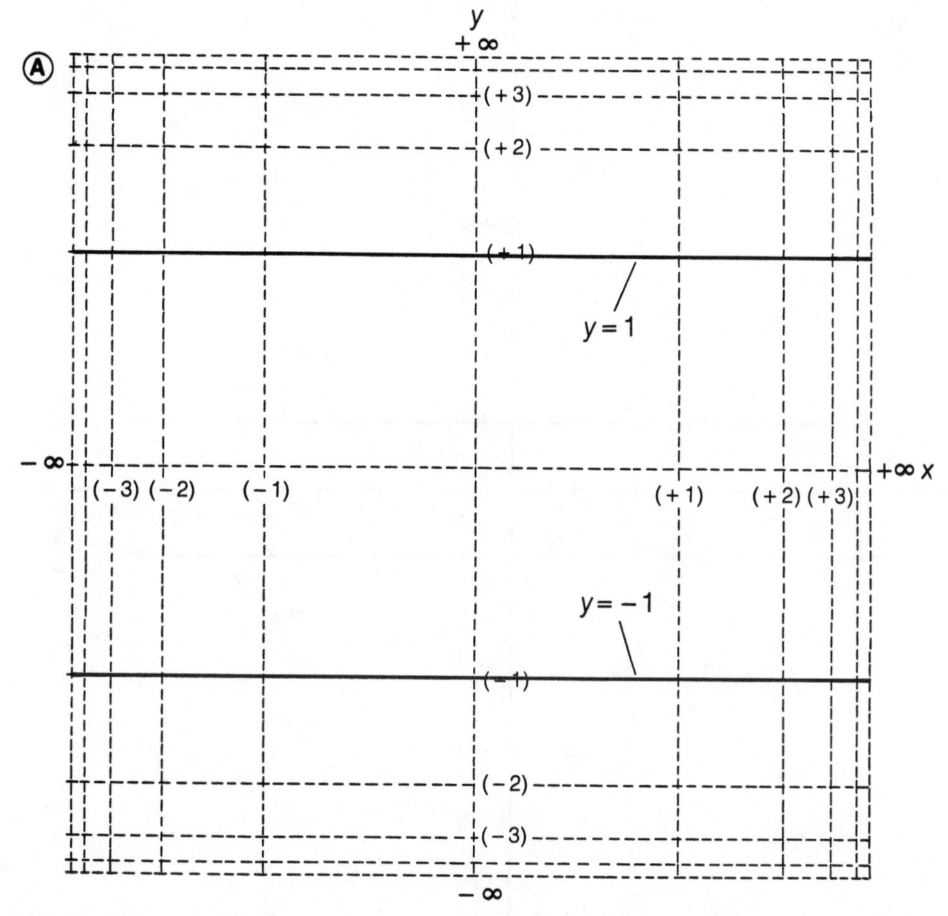

Fig. 6-3. At (A), parallel lines $y = 1$ and $y = -1$ do not appear to intersect on the IC plane, even at the infinite extremes.

ordered pairs and triples of real numbers can account for every point on a plane or in three-space. This question is involved with the Continuum Hypothesis. It has been shown that the Continuum Hypothesis can be assumed to be true or false, with contradictions apparently not arising in either case. Whether this hypothesis is assumed true (the real numbers can be paired off with *all* of the points on the geometric line) or not, we do get the result that the set of points that does correspond to the real numbers on a line with linear demarcation of integers is an infinitely long line. It has no beginning and no end, and one never repeats the passing of a given point once it has been passed in a specific direction.

Nonetheless, we know that we can pair off real numbers with finite line segments as we did in the previous chapter. It is not necessary that a line be infinitely long in order to accommodate all of the real numbers. Nor is it necessary that a plane have infinite expanse, or that space be infinitely large, in order for all possible ordered pairs and triples

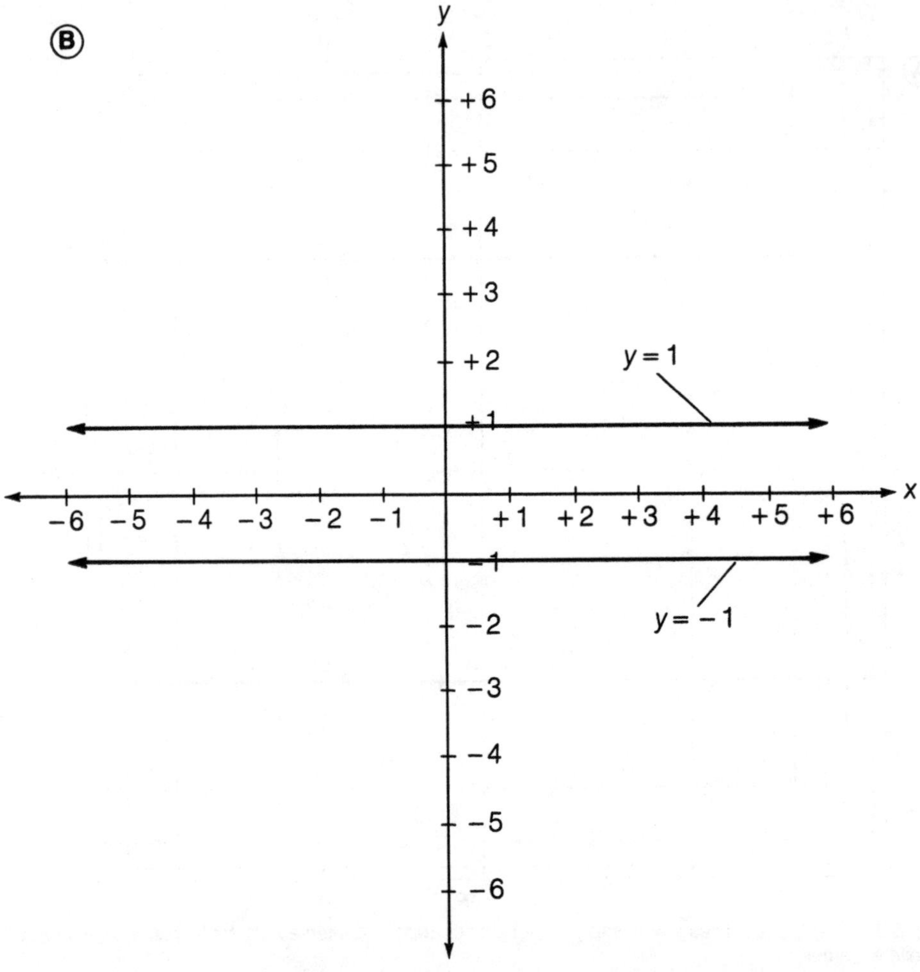

Fig. 6-3. (B) shows the two lines appear parallel also on the Cartesian plane (B).

to be represented. Infinite space is therefore *not* implied by the existence of infinitely many, densely packed real numbers. It may well be that there is no such thing as truly infinite space, or infinite planes or lines. In the four-sphere universe, and in any finite-but-unbounded, n-dimensional universe, there is no such thing as an infinite Euclidean line, plane, or space.

We imagine such infinite geometric objects because, perhaps, we look for some excuse to get away from the intuitive notion that geometric infinities might not exist on the large scale. Since Isaac Newton's time, we have been conditioned to believe in infinite space, and such long-standing beliefs are hard to give up.

If space is in fact finite, then Euclidean postulates fall apart, and we may have such odd triangles as 30-90-90 or 90-90-90 on the surface of a four-sphere. Imagining two of the vertices at the equator of a globe and one vertex at the North Pole makes it possible to visualize such strange triangles. We may even have a 180-180-180 triangle, with all three of the vertices lying on the equator.

For purposes of discussion, we have allowed for infinite space in the analysis of parallel lines intersecting with 0-degree angles; but in a finite-but-unbounded space, there is no need for any of this because there are actually no parallel, straight lines. It might be possible for two lines to exist that never intersect; imagine the equator and the Tropic of Cancer, for example. But one of the lines—in this case the Tropic of Cancer—is not straight in the sense that it is not a great circle or geodesic on the sphere. On the surface of a four-sphere, this would also hold true: if two lines never intersect, then one or both must not be truly straight in the sense that one or both lines do not represent the shortest possible distance between two points. In our universe, we consider that photons, the particles of radiant energy, will always travel in straight lines (in the geodesic sense) in a vacuum. The apparent bending of light rays by a gravitational field is simply the result of distortion in space, causing the geodesic to be different than it would be if the gravitational source were absent.

Extreme Trigonometric Values

We return again to the concept of Euclidean, infinite space, realizing that it represents an ideal that might not actually exist.

You ought to be familiar with the definitions of trigonometric functions, as represented by right triangles and by the unit circle. For a given right triangle *ABC* (Fig. 6-4), having opposite sides of lengths *a*, *b*, and *c*, we may denote the trigonometric functions as

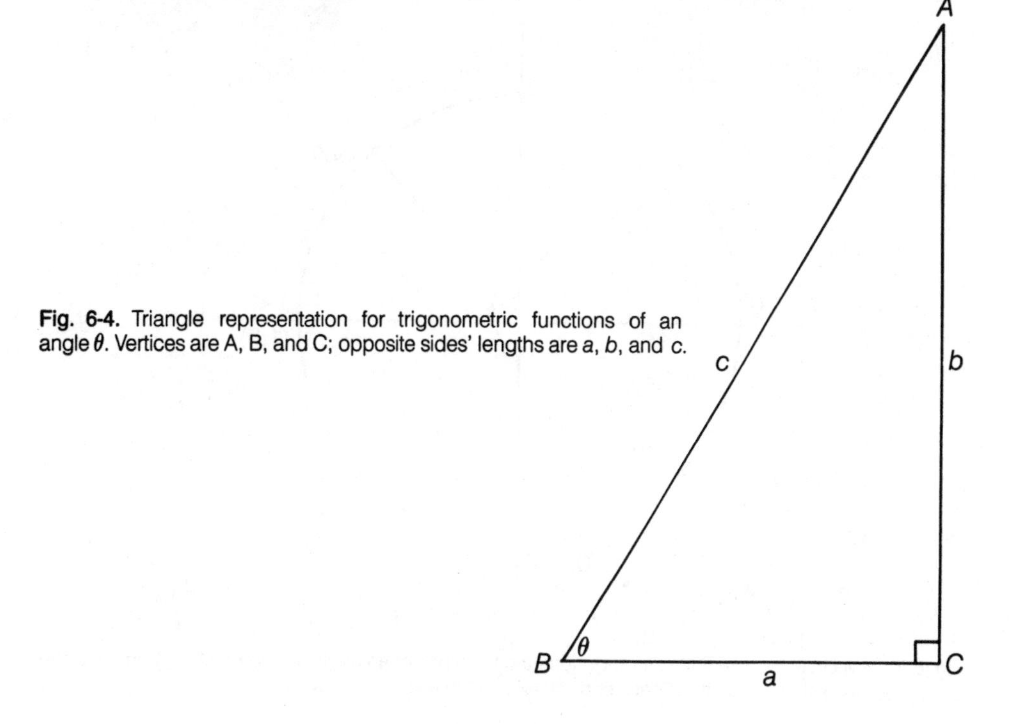

Fig. 6-4. Triangle representation for trigonometric functions of an angle θ. Vertices are A, B, and C; opposite sides' lengths are *a*, *b*, and *c*.

follows for the angle θ:

$$\text{sine function} = \sin\theta = b/c$$
$$\text{cosine function} = \cos\theta = a/c$$
$$\text{tangent function} = \tan\theta = b/a$$
$$\text{cosecant function} = \csc\theta = c/b$$
$$\text{secant function} = \sec\theta = c/a$$
$$\text{cotangent function} = \cot\theta = a/b$$

The unit-circle representation is shown in Fig. 6-5. Here, for a given angle θ at the origin with values (x,y) on the circle:

$$\sin\theta = y$$
$$\cos\theta = x$$
$$\tan\theta = y/x$$
$$\csc\theta = 1/y$$
$$\sec\theta = 1/x$$
$$\cot\theta = x/y$$

Certain trigonometric functions are not defined when the angles are of certain values. The triangle method of Fig. 6-4 applies only for angles that are greater than zero and less

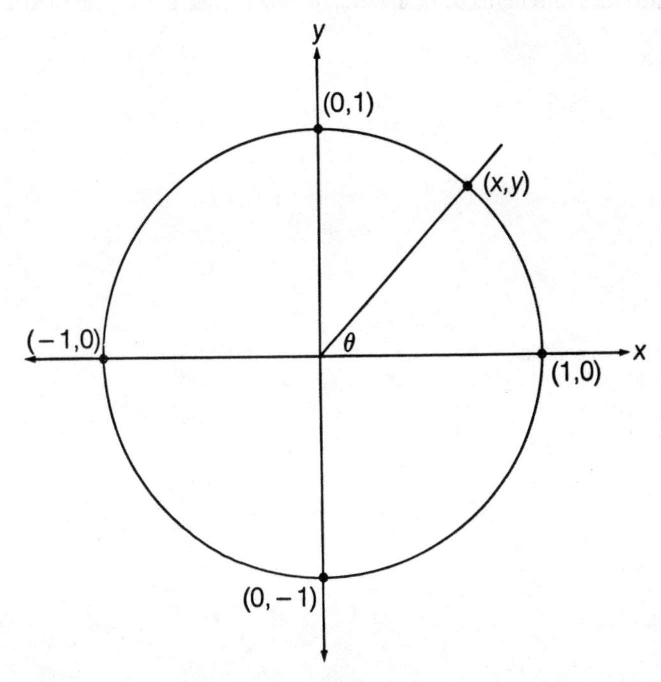

Fig. 6-5. Unit-circle representation for trigonometric functions of an angle θ. The point (x,y) on the circle determines the values of the functions, as discussed in the text.

than 90 degrees. The circle model of Fig. 6-5 appears to apply for any angle, even negative angles and angles measuring more than 360 degrees, with the exception of certain angles where a quantity might have zero in the denominator. Here, it is precisely these extreme angles that interest us.

Let us first consider the situation when $\theta = 90$ degrees. This is shown in the diagram of Fig. 6-6: a "triangle" with two sides that are, apparently, rays having infinite length. Let us arbitrarily say that the lengths of the sides b and c are infinite and denote them as having measures of aleph-null units ($\aleph_0$). How can we say this? If we mark the sides of the triangle off in units as if we were constructing a number line, making the length of side a (the base) equal to one unit, we can keep on counting these units for sides b and c, and this makes the lengths denumerably infinite in terms of the base side a. We will use our intuition, first, to attempt to determine the trigonometric function values. It would seem that:

$$\sin 90 = b/c = \aleph_0/\aleph_0$$
$$\cos 90 = a/c = 1/\aleph_0 = 0$$
$$\tan 90 = b/a = \aleph/1 = \aleph_0$$
$$\csc 90 = c/b = \aleph_0/\aleph_0$$
$$\sec 90 = c/a = \aleph_0/1 = \aleph_0$$
$$\cot 90 = a/b = 1/\aleph_0 = 0$$

We do not yet know specifically what to call the value $\aleph_0/\aleph_0$. It is intuitively apparent that $\aleph_0/1 = \aleph_0$, and also that $1/\aleph_0 = 0$. (We do not yet acknowledge the possibility that there may be infinitesimals—little numbers smaller than any of the reciprocals of natural numbers, yet bigger than zero. Their existence has been considered, however.)

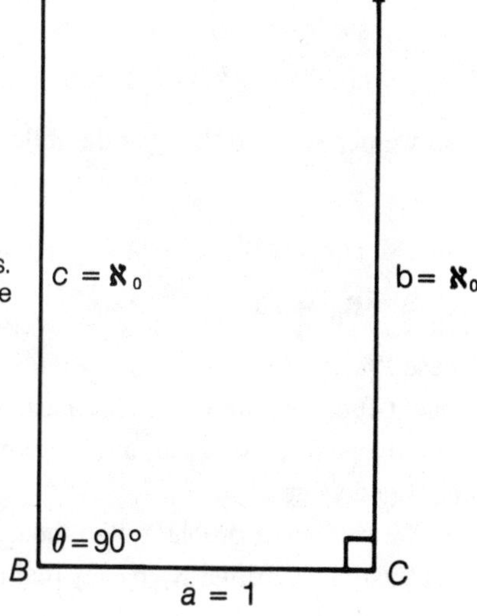

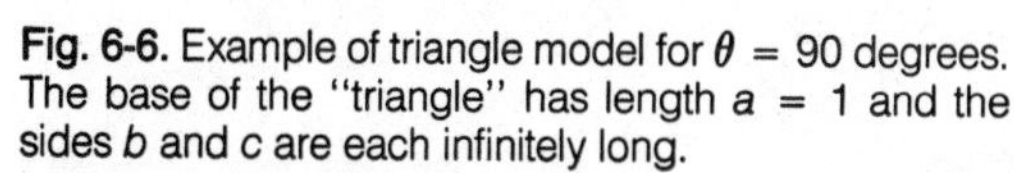
Fig. 6-6. Example of triangle model for $\theta = 90$ degrees. The base of the "triangle" has length $a = 1$ and the sides b and c are each infinitely long.

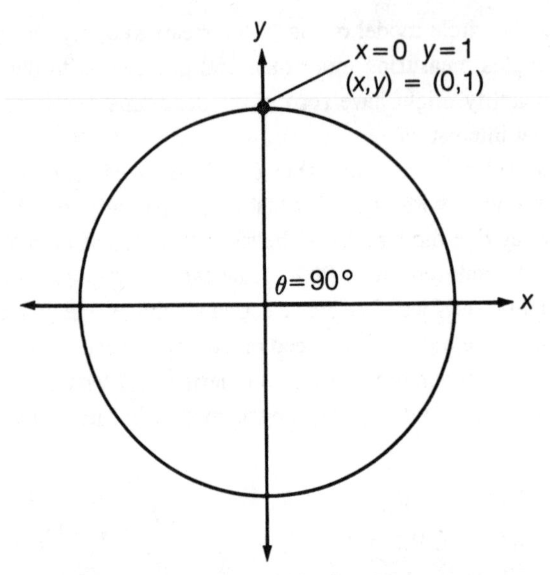

Fig. 6-7. Unit-circle model for θ = 90 degrees.

Now let us observe the situation from the standpoint of the unit-circle model. Here, we needn't even draw a diagram, although I have done so for clarification. The point in question is $(x,y) = (0,1)$, the 90-degree point on the unit circle $x^2 + y^2 = 1$. This is shown in Fig. 6-7. The values of the trigonometric functions are then determined as:

$$\sin 90 = y = 1$$
$$\cos 90 = x = 0$$
$$\tan 90 = y/x = 1/0$$
$$\csc 90 = 1/y = 1/1 = 1$$
$$\sec 90 = 1/x = 1/0$$
$$\cot 90 = x/y = 0/1 = 0$$

Can we derive from this that the following hold?

$$\aleph_0/\aleph_0 = 1$$
$$1/\aleph_0 = 0$$
$$\aleph_0 = 1/0$$

These "facts" seem to follow when we compare the two models of the sine, cosine, and tangent functions and their inverses. Certainly the functions ought to mean exactly the same things in either case, and we would like to think that this should hold true even for undefined values.

We run into a problem with the quotient $\aleph_0/\aleph_0 = 1$. We know that we can multiply the transfinite cardinal $\aleph_0$ by any real number (except zero) and obtain either $\aleph_0$ or $-\aleph_0$.

Specifically, if we say that $2\aleph_0 = \aleph_0$, then we may say that $2\aleph_0/\aleph_0 = 2(\aleph_0/\aleph_0) = 2$; but also that $2\aleph_0/\aleph_0 = \aleph_0/\aleph_0 = 1$. Therefore $2 = 1$, which is absurd.

We get around this problem by introducing the concept of *contingent equality* between *numerical entities*. The rudimentary principles of entity theory are discussed in *Puzzles, Paradoxes and Brain Teasers* (TAB Book No. 2895, 1988). If we consider that the quotient $\aleph_0/\aleph_0$ may take all of the positive real-number values, and if we consider that contingent equality is not transitive, we get rid of this contradiction.

All of this is nonrigorous. There was a mathematics professor who once told me that "division by zero" was absolutely meaningless, but that I could consider the ratio operation and symbolize it with the slant bar (/) or the colon (:) or the horizontal line for longer equations (—). Then I could formulate axioms such as the following.

AXIOM. For any numerical entities a, b, and c, $a(b/c) = (ab)/c = b(a/c)$.

I would prefer to use the Greek lower-case alpha, beta, and gamma for the entities, but they are hard to include in typeset form, and they also tend to intimidate readers who don't know what they are (and therefore pronounce them "huh," "huh," "huh" respectively).

What is the significance of all of this? Why should we care what the result of having a zero in the denominator of a fraction might be? It is perhaps a desire to have a "closed" system of arithmetic, where there is no combination of numbers and operations that cannot be defined or that has no meaning.

In *Puzzles, Paradoxes and Brain Teasers*, I demonstrated a method of "proving" that the infinite value, or supposed infinite value, that results from "division by zero" can actually be considered to be zero itself. We do not seem to get a contradiction when we allow this to be the case. The tangent of 90 degrees would then take on the value zero. If we graph the tangent function with the values tan (90) = tan (270) = 0, we obtain a function that is not continuous, but nonetheless is defined for every value in its domain (Fig. 6-8A). We can also graph such functions as $y = 1/x$, and have a value for $x = 0$ at the origin (0,0), as shown in Fig. 6-8B. This way of "defining" ratios having denominators of zero seems, however, to be a kind of cop-out. I believed, when I found that I could prove this about the numerical entity $I = 1/0$, that I had made the axioms too strong, and that it was entirely possible that the whole theory would eventually run into a contradiction and prove its worthlessness. I didn't pursue the matter too much further, but it seemed that weakening the axioms made the system must less strong and practically vacuous. I have yet to find a good theory that appears sound and has substantive content for defining 1/0. It is perhaps exactly this sort of frustration that has discouraged mathematicians from pursuing infinity much along these lines. There are other, geometric methods of evaluating the infinite that provide interesting results corresponding with established theorems.

The Paradox of Concentric Circles

How many points are there around the circumference of a circle? If we take the point of view of conventional geometricians, we will answer, "infinitely many." There is the Continuum Hypothesis that states that the number of the continuum, and thus the number

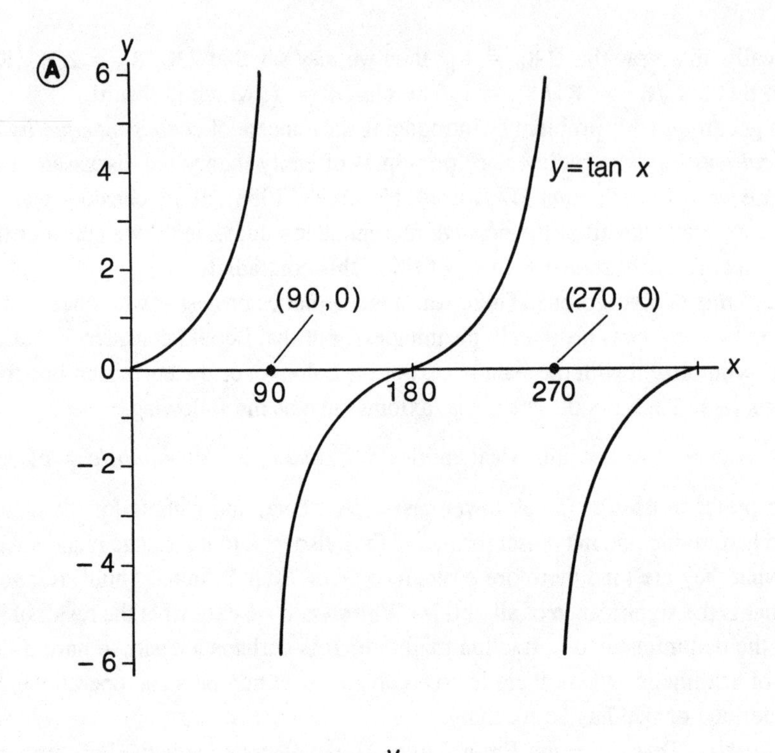

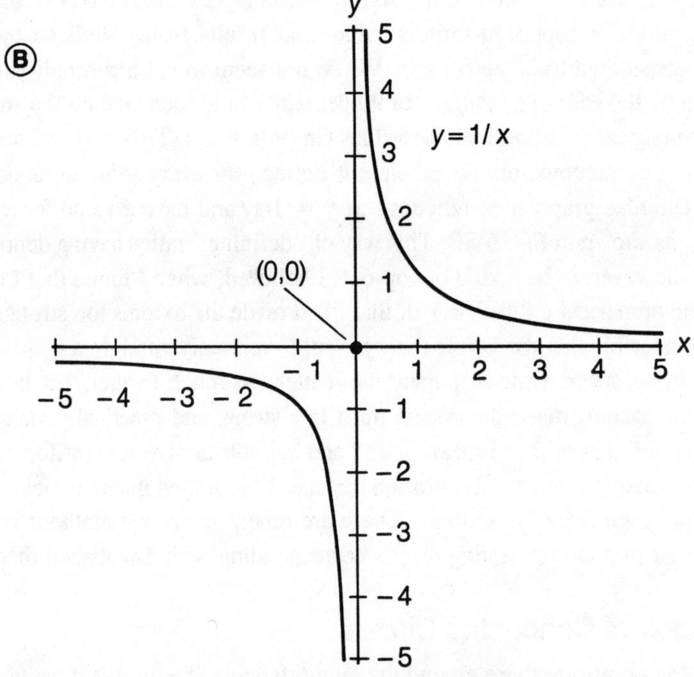

Fig. 6-8. (A) represents the tangent function with angles in degrees from 0 to 360. The points (90,0) and (270,0) have been added. (B) shows the function $y = 1/x$, with (0,0) added.

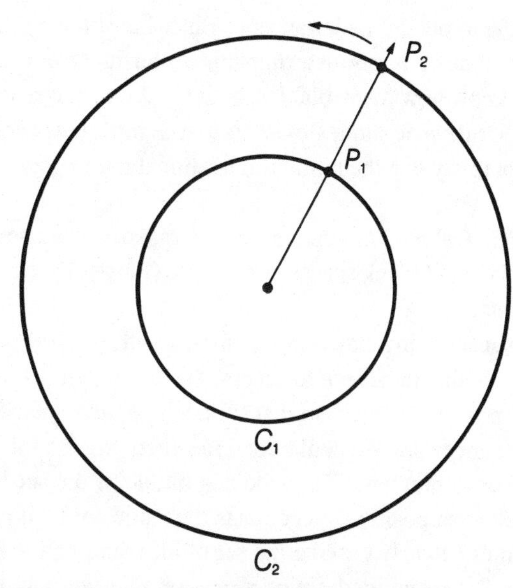

Fig. 6-9. Two circles with a common center and one with a radius twice the other have an identical quantity of points in their circumferences. The two circles are denoted C_1 and C_2, and specific points on the circles that intersect a given radial are denoted P_1 and P_2.

of points in the circumference of any circle, is aleph-one ($\aleph_1$). We may simply regard the number as "infinity" and denote it by the lemniscate ∞.

In Fig. 6-9, we have two concentric circles, one with twice the radius (and hence twice the circumference) of the other. How many points are there in the circumference of the larger circle? We answer, "infinity." The same answer is given when we ask the number of points along the circumference of the smaller circle. We can in fact demonstrate that there exists a one-to-one correspondence between the points in each circle, by drawing a radius line from the common center point through both circles. As we sweep this radius line around a full 360 degrees, we note that, for each intersection point of the radius with the smaller circle, there is one point where it intersects the larger circle. The converse is also true: for each point where the radius intersects the larger circle, it also intersects the smaller circle at one point.

How can we make this dovetail with the notion that there ought to be twice as many points around the larger circle because its circumference is twice as great?

Galileo Galilei, of astronomical fame, had an answer to this apparent paradox. He proposed that an infinitude of infinitely small spaces could be interposed between the points of the smaller circle, obtaining the larger one with the same number of actual points.

However, we might counter by constructing a pair of circles having a random, or unknown, ratio of radii, and the same would hold. How could we interpose blank spaces between points by any system without knowing the ratio of blank spaces to add relative to the number of points? If the two circles had integral radial sizes, for example 45 and 67, then we could say that we would have to add just $67 - 45 = 22$ blanks for every 45 points in the smaller circle to obtain the larger one. But what if the circles were in ratio such as $1:\sqrt{2}$? The latter number is not determinable in terms of any fraction of integers. We would simply not know how many blank spaces to add. It may also be argued that any two randomly chosen concentric circles exist in a size ratio that cannot be determined precisely at all, except to say that the ratio is an irrational number and hence not exactly determinable.

In this manner, Galileo's argument cannot make any sense. It is just as well—the thought of infinitely small blank spaces bothers me (for one) even more than the thought of infinitely tiny points.

There is a corresponding situation that arises with integers and all countable infinite sets, and it is easy to illustrate if not to accept. We could pair off the even positive integers with the set of all positive integers so that for each positive integer n, there exists exactly one even positive integer $2n$. We could continue each process for a countably (denumerably) infinite number of numbers. Thus the magnitude of the set P of all positive integers and the set Q of all even positive integers has the same cardinal number, which we know now to be aleph-null ($\aleph_0$). If we take the set of all reals, call it R, and then double each real in R to obtain S, we end up defining a new set that not only has the same cardinality (believed to be aleph-one or $\aleph_1$) but is in fact the very same set! For every real number r in R, there exists a number $s = 2r$ in S; and for every s in S there is exactly one $r = s/2$ in R. This is exactly the situation that is illustrated by two concentric circles, one with twice the circumference of the other, as in the drawing.

Infinigons

It is possible for a figure to have infinitely many sides, at least in the mindscape. An example of this is the Koch curve.

To form a Koch curve, begin with an equilateral triangle, as shown in Fig. 6-10A. Then trisect each side of this triangle, using the middle third of each side as the base for a new equilateral triangle. When you have constructed the new equilateral triangles, erase the bases to obtain a figure like that shown in Fig. 6-10B. This process is then repeated; each side of the new figure is trisected, and the middle third of each side is used as the base for a new equilateral triangle. Then the bases of the new triangles are erased, yielding a polygon like that in Fig. 6-10C. This process is then repeated aleph-null times: it is done over and over, without end.

It is clear that the final figure, the Koch curve, cannot have finitely many sides. We increase the number of sides with each step.

You may, if you wish, begin with a very large equilateral triangle and work for several hours drawing a very precise replica of the Koch curve to so many iterations that further detail is not possible with the fineness of the point of the marker you are using. You can begin to see that an actual Koch curve would look very similar to what you have drawn. You can step a few feet or yards away from your drawing and imagine that it is a true Koch curve in all its infinite detail. We can conveniently neglect the fact that the line would have

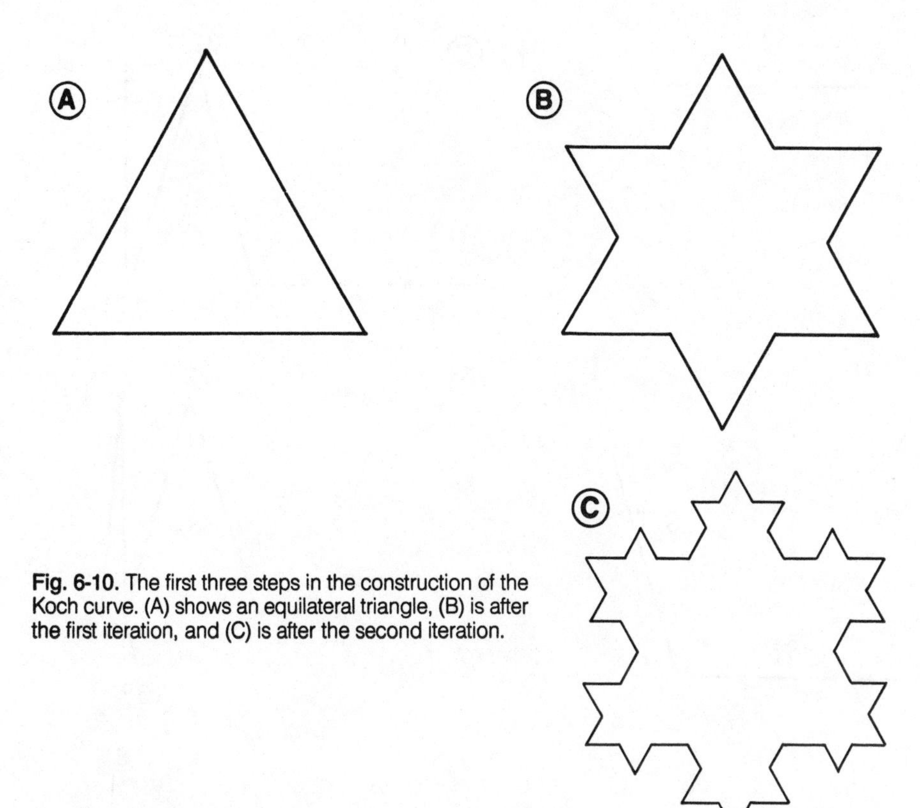

Fig. 6-10. The first three steps in the construction of the Koch curve. (A) shows an equilateral triangle, (B) is after the first iteration, and (C) is after the second iteration.

to be infinitely fine, a true geometric line in the sense that Euclid surely imagined a line. It is a real figure in the mathematical mindscape.

The Koch curve is named for Helge von Koch (pronounced "Cook"), the Swedish mathematician who is thought to have been the first to describe it. There are other infinite-sided polygons, or "infinigons," that can be imagined.

Clearly the Koch curve has finite area. We can circumscribe it with a circle, knowing that the area of the Koch curve must thus be less than the area of the circumscribed circle. Another thing we can say about the Koch curve is that in a sense, it is continuous (a line that is not broken). It resembles the shoreline of a lake or an island—the detail is always there and always looks quite similar at all scales—and is extreme in magnitude. We could never work our way completely around the Koch curve and make a finite number of moves. The measure of the interior angles of this curve is infinity, greater than any finite number.

Another method of construction an "infinigon" is shown in Fig. 6-11. We begin at (A) with a line segment measuring one unit in length. We then draw a line segment of length one unit, going upward at an angle from the first line segment (B). We now have sides S_1 and S_2, as shown. The angle subtended between sides S_1 and S_2 is something greater than 90 degrees but less than 180 degrees; let us call it 90 + 16 or 106 degrees. (I have chosen that value for a reason that will become clear momentarily.) Now we draw

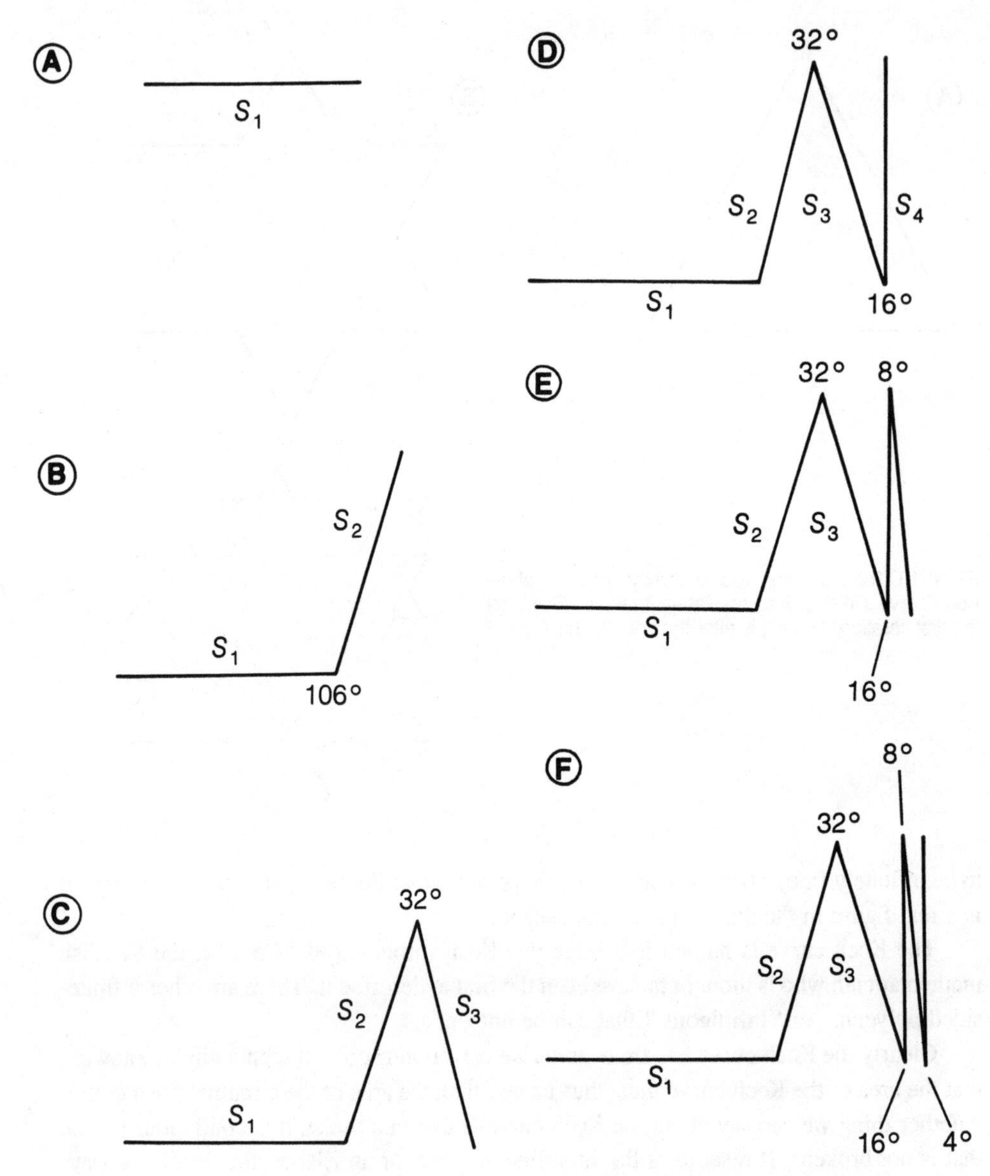

Fig. 6-11. Construction of an infinite-sided polygon. At (A), we begin with a line segment S_1. At (B), we construct S_2 at an obtuse angle relative to S_1 and the same length as S_1. Succeeding steps at C, D, E, and F are discussed in the text.

side S_3 having one unit, making S_2 and S_3 the equal-length sides of an isosceles triangle whose base is not drawn. This means that the angle between S_2 and S_3 is equal to 32 degrees in this case (Fig. 6-11C).

We now draw sides S_4, S_5, S_6, and so on, each side having one unit of length but each side subtending an angle of just half the previous angle. Thus, the angle between S_3 and S_4 in Fig. 6-11D is 16 degrees; the angle between S_4 and S_5 in Fig. 6-11E is 8 degrees; the

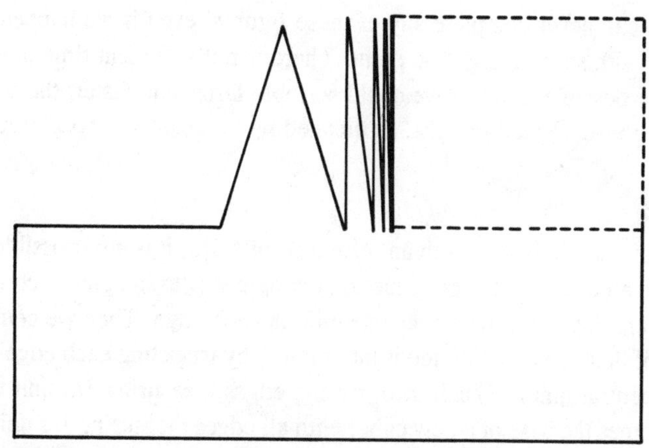

Fig. 6-12. The infinigon when complete. Note that we cannot say where the last line segment ends (as indicated by dotted lines completing the figure) because there are infinitely many of them.

angle between S_5 and S_6 in Fig. 6-11F is 4 degrees. And so it continues: the angles are halved each time, and it becomes possible to cram an infinite number of line segments into a finite space.

We then draw the last three sides of what appears now to be a serrated rectangular blade of some sort, obtaining the irregular infinigon of Fig. 6-12. This figure has infinite perimeter, an infinite number of sides, and an infinite measure to the sum of all its interior angles. It has a finite area, however, because it can be circumscribed.

Now imagine drawing this figure, but each time halve the lengths of the sides so that $S_1 = 2S_2 = 4S_3 = 8S_4 = \ldots$. When the figure is complete, there is only a finite perimeter, a finite area, and an infinite number of sides; the sum of the measures of the interior angles will be infinite. This is shown in Fig. 6-13.

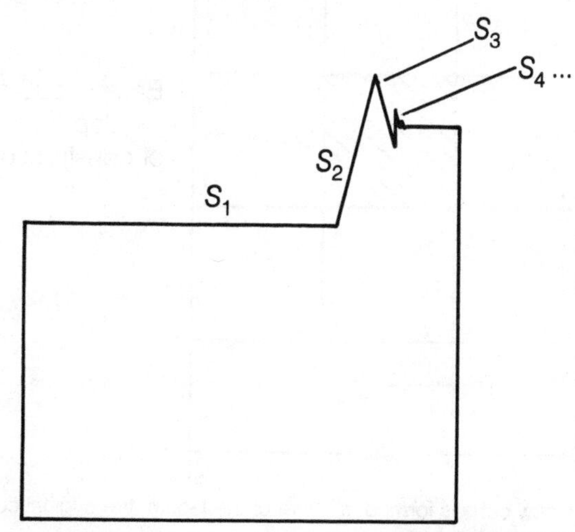

Fig. 6-13. An infinigon with definite completion line segments. The sides S_1, S_2, S_3...are halved in length each time they are drawn. This figure can be shown to have a finite perimeter, unlike the polygon in Fig. 6-12, which has an infinite perimeter.

Actually, we can never complete any of these figures, even if we happen to obtain a Euclidean pencil with an infinitely fine point. There is not sufficient time in the real cosmos for this to be possible. Even if we could scribble faster and faster, there is a limit to how fast we can go—the speed of light, as dictated by the results of special relativity.

Infinihedrons

Just as we can have polygons with an infinitude of sides, it is also possible to imagine polyhedrons with an infinite number of faces. A simple way to imagine such a figure is to think of a cube, and say that it measures one unit on each edge. Then we could take each of the six faces of this cube, and divide it into ninths, by trisecting each edge as shown in Fig. 6-14. The central ninth of each face, having edges measuring $^1/_3$ unit and area $^1/_9$ square unit, becomes the base of a new cube, with all edges measuring $^1/_3$ unit and consequently a volume of $^1/_{27}$ or $(^1/_3)^3$ cubic unit.

This process is repeated, just as was done with the Koch curve. The result will be an object that can be completely surrounded by a sphere, so it will have a finite volume. But it will have infinitely many faces. I have not attempted to illustrate such an object, even after the very first step following the cube, simply because a drawing would be too difficult to construct and would not do justice to the object.

Any of the regular polyhedrons can be expanded upon in this way, making what we might call *regular infinihedrons*. There is nothing special about taking $^1/_3$ of the value for edges of the polyhedron; we might take any fractional value less than 1. In general, all such regular infinihedrons have finite volumes.

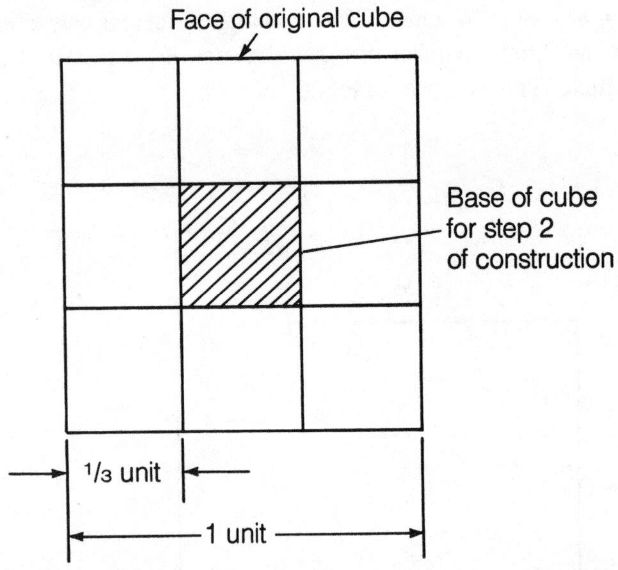

Fig. 6-14. The base of a new cube is formed from $^1/_9$ of the face of the original cube in the process of constructing an infinihedron using cubes.

Infinite Angles

You may have heard about angles measuring more than 360 degrees or less than zero. Usually, such angles are reduced to some value greater than or equal to zero degrees and less than 360 degrees. This may be done by subtracting or adding multiples of 360 until the angle falls within the range of $0-360$ degrees or $0-2\pi$ radians.

Rarely do we encounter such angles unless we are involved in an analysis of some property or function that involves trigonometry. The trigonometric functions allow for angles to go indefinitely into the positive and negative directions so that the periodic function might repeat many times. What do we do if we have an angle specified as having a measure of $\aleph_0$ degrees or the negative of this value?

It is fairly evident that we could take an angle of infinite size and regard it as having any measure at all or all possible measures from zero to 360 degrees simultaneously. We might define such an angle, call it ϕ, as a numerical entity having values such that $0 \leq \phi < 360$ if we use degrees and $0 \leq \phi < 2\pi$ if we use radians. This might seem rather silly, but it is no sillier than the analysis of any other property of aleph-null.

The situation for the sine function is shown at Fig. 6-15. At (A), the function is graphed on the infinite coordinate (IC) plane. AT (B), we "zoom" in on the positive axis of the IC plane so that we can get a closer view of values greater than 360 degrees. No matter how much we "magnify" the axis, we will always end up with a compressed sine wave that appears more or less the same. At (C) in the figure, the sine function is shown in Cartesian coordinates so that several cycles repeat.

The value of the sine function at the infinity points (positive and negative) appears to take all of the values from -1 to $+1$ inclusive. We would expect this for an angle that could take any value whatsoever: the numerical entity resulting from taking $\sin \phi$ should be equal to the entity comprised of all the values $\sin \theta$, where θ is some angle and $0 \leq \theta < 360$ degrees.

We can do similar graphs with the other trigonometric functions. In the case of the tangent function, we may define $\tan 90°$ as either $\aleph_0$ and $-\aleph_0$ together—that is, an entity consisting of the transfinite cardinal $_0$ and its negative—or we may consider $\tan 90°$ to be zero. If we do the latter, the infinity points (positive and negative) will not appear in the graph for $y = \tan 90x$. But the entire range of real numbers will appear in any case. The value $\tan \aleph_0$ includes all real numbers in an entity that we might call R.

The Binary Tree

It has been demonstrated, in general, that if we take the number 2 and raise it to any power x, then the value 2^x must be greater than x, even if x happens to be a transfinite cardinal. Thus it follows that $2^{\aleph_0}$ must be larger than aleph-null. It is interesting to think that we can depict the number of points on a line segment by means of a device called the binary tree, illustrated in Fig. 6-16, as $2^{\aleph_0}$.

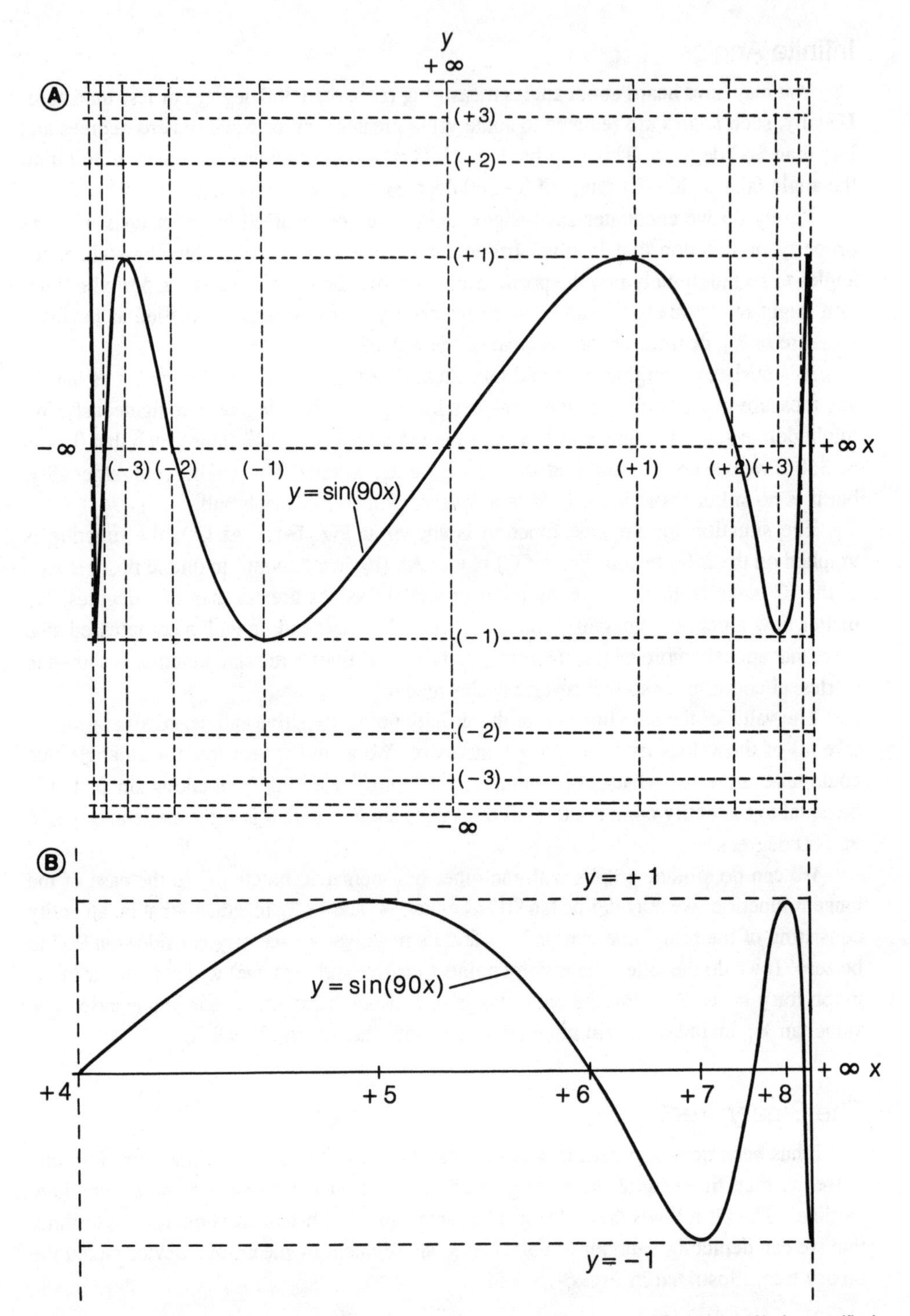

Fig. 6-15. The function $y = \sin(90x)$, for angles in degrees, is graphed on the IC plane at (A). A magnified view along the x axis is shown at (B).

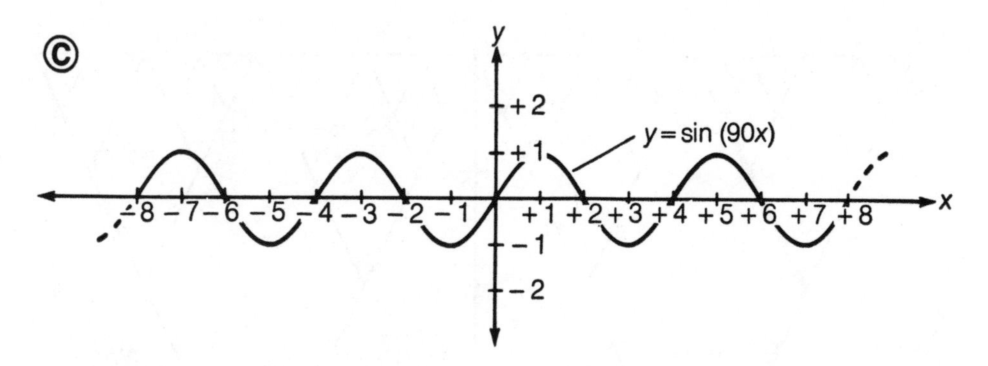

Fig. 6-15. At (C), the function is shown graphed on the Cartesian plane. Note that at the infinity and minus-infinity values of x, the function appears to have all values y such that $-1 \le y \le +1$.

The binary tree is a curious object because it increases in number of branches much faster than the position one goes up to, as indicated by the level numbers 1, 2, 3, 4, and so on in the drawing. Actually, for level n, there are 2^n branches in the binary tree. At the very bottom where $n = 1$, there exist $2^1 = 2$ branches. At the level where $n = 2$, there are $2^2 = 4$ branches; at $n = 3$ there are $2^3 = 8$ branches. The number of branches that we may choose doubles each time we go up one more level. Yet, note that the levels are squashed in exactly the same way as the axes of the IC plane are squashed. Therefore, there will be $\aleph_0$ levels in the binary tree before we reach the line segment at the top. At this level, $n = \aleph_0$, there will be $2^{\aleph_0}$ branches, and this has been proven to be greater than $\aleph_0$.

It is tempting to suppose that the branches of this binary tree will cover every point along the line segment at the top of the tree (the aleph-null level). But this depends on whether we think that the *nodes*, or branch points, occupy the line segment or whether branches occupy it. We cannot tell by looking at the drawing nor by magnifying it to any finite extent. Does the tree "end" with node points or with branches? We might say either one! Therefore, while it appears that there may be $2^{\aleph_0}$ points on the line and that the number of the continuum must therefore be $2^{\aleph_0}$, we do not really know for certain that this is the case based on this drawing.

It is still odd to think that the total number of nodes that we will encounter on any trip up the tree is aleph-null, but there are more than this cardinal number of branches we might travel through. After all, aren't there just two branches coming off every node point? Yet the demonstration is clear: there are $2^{\aleph_0}$ branches, and $2^{\aleph_0} > \aleph_0$.

It is clear that there must be room for all of the branches to end at the line segment. Thus, there are at least $2^{\aleph_0}$ points on the segment. But there could be more; we cannot say from this alone.

The Continuum Hypothesis, as Georg Cantor believed it, says that $2^{\aleph_0} = \aleph_1$, and that this is indeed the exact number of points on a line segment (the so-called number of the continuum). But the Continuum Hypothesis is undecidable in set theory. We do not seem to have any problem assuming it to be true; this does not lead to a contradiction. But we can do just as well by assuming it is not true.

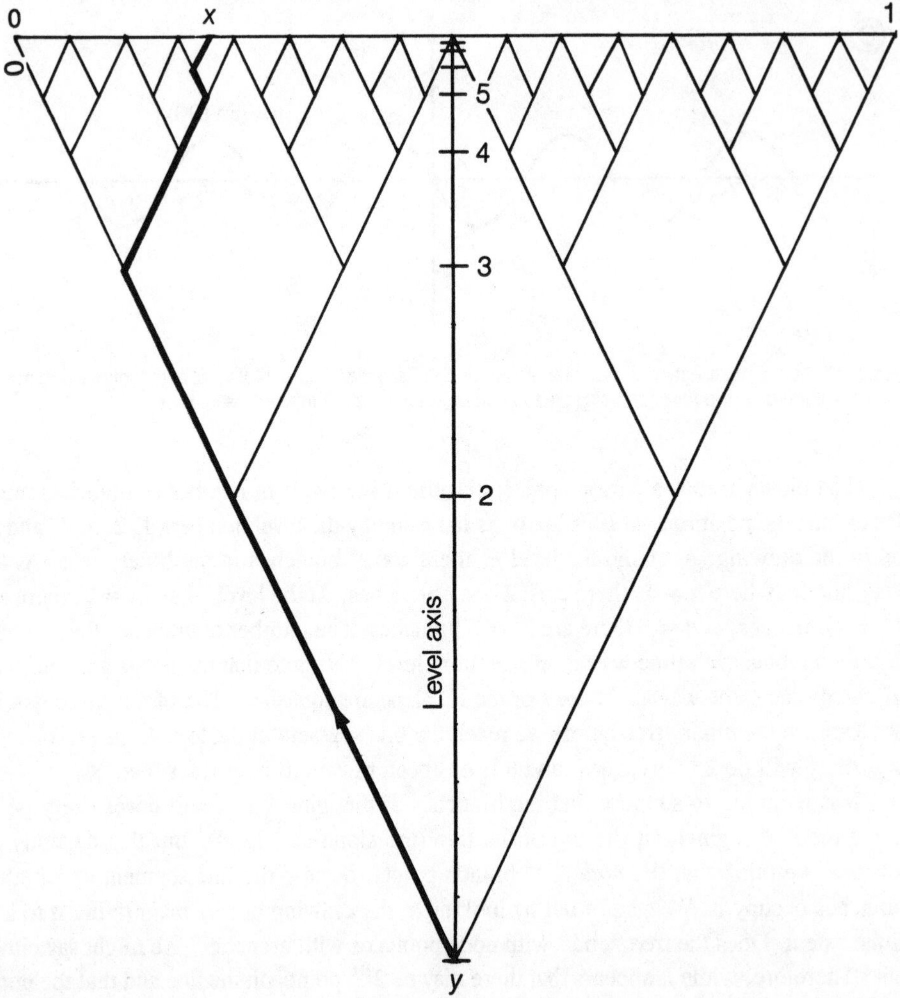

Fig. 6-16. The binary tree. Levels are shown on the vertical line segment at the center (not itself a part of the binary tree). The line segment at the top represents the end of the tree. Here, we show the tree constructed up through level $n = 5$. The heavy line depicts the path to arrive at a point x on the top line segment, where $0 \leq x \leq 1$. Does the binary tree account for all real numbers and all points on the line segment between 0 and 1 inclusive?

In actual space, then, we are driven to postulate the number of points that there are. Proving that the number is aleph-one or that it is something greater than aleph-one is evidently not possible either way. We do not have, with the current axioms of set theory, a geometric microscope powerful enough to resolve the question of how many nothings it takes to make something. And that is what we are asking when we inquire as to the number of *points* that there are in a certain linear span, area, or volume with finite measure. It should not come as any great surprise that the answer to such a meaningless question, based on concepts taken as elementary and not to be defined, should be undecidable. The real surprise would be if the question *did* have a specific answer.

Further Reading

Bunch, Bryan H. *Mathematical Fallacies and Paradoxes*. Van Nostrand Reinhold Company, 1982.

Gamow, George. *One, Two, Three…Infinity*. Viking Press, 1947.

Gibilisco, Stan. *Puzzles, Paradoxes and Brain Teasers*. TAB BOOKS, 1988.

Guillen, Michael. *Bridges to Infinity*. Jeremy P. Tarcher, Inc., 1983.

Rucker, Rudy. *Infinity and the Mind*. Bantam Books, 1983.

Index

Q

R

S

T

U

V

Z